SI METRIC UNITS

An Introduction

Revised Edition

SI METRIC UNITS

An Introduction

Revised Edition

H. F. R. ADAMS

Chief Instructor, Electronics Division
British Columbia Vocational School
Burnaby, B.C.

McGraw-Hill Ryerson Limited

Toronto Montreal New York London Sydney
Johannesburg Mexico Panama Düsseldorf
Singapore Sao Paulo Kuala Lumpur New Delhi

SI METRIC UNITS: An Introduction

Revised Edition 1974

2 3 4 5 6 7 8 9 MP 10 9 8 7 6 5 4

Printed and bound in Canada

Table of Contents

Contents

Introduction

In 1965, the Government of the United Kingdom accepted the proposals of British industry, and decreed that the country would convert to the metric system of measurements by 1975.

In 1968, the United States Government set up a committee to report on the possible use of the metric system as the only official measuring system for the U.S.A.

In 1970, the Government of Canada issued a White Paper which concluded that the eventual adoption of the metric system should be an objective of Canadian policy.

When these three countries eventually complete their conversions to the metric system, and, more specifically, to that form of metric known as SI, then virtually every manufacturing nation in the world will be using the same standards of length, mass, time, electric current, temperature, light intensity and molecular substance. In addition, they will all be using the same powers-of-ten multiples and submultiples. There will be a new era in simplified communication.

The conversion in North America will not take place overnight. Canadian and American scientists already use the metric system, and technologists and technicians in some fields, such as medicine and electronics, are very familiar with many, if not all, of the SI units.

This book has been written to help such people as

teachers
technicians-in-training
technicians in industry
tradesmen
office workers
and the general public

to learn *as much as they need* about the metric system during the early stages of the North American conversion to metric.

The author has studied in depth the manner in which the conversion was effected in the United Kingdom, and his presentation is colored by the British reports and observations to which he has had access. From this history, several points must be emphasized at the outset of the book:

1 Everyone should learn enough to have a working understanding of the *total concept* of the International System of units. (Think "system".)
2 Everyone should develop, as quickly as possible, some *mental associations* with metric units. (Think metric.)

3 The best time for "old timers" to learn metric units *in depth* is just before they are going to be needed — on the job, in the shop, or in the home.
4 An "old timer" should *concentrate* only on those metric units which he will need immediately.
5 The best way for an "old timer" to learn metric is to *form lasting mental associations* in metric units, rather than repeatedly converting the new metric units to the old imperial units.

Here, of course, an old timer is any person, of any age, who has been schooled in imperial units, works in imperial units, or teaches in imperial units.

Because any measuring system involves making measurements, and performing calculations with those measurements, this book may appear at first to be a mathematics textbook. This is not the case, though. My main purpose is to help you to prepare for the psychological impact of making and working with measurements in a system which is, in whole or in part, unfamiliar to you.

Since people in all walks of life, with widely varying backgrounds, will be using this book, it has been difficult to provide suitably graded examples which will enable some exercises at least to be meaningful to every reader. It has been assumed that you are familiar with the following *ideas*, even if you may not be proficient in their use:

concept of decimal numbers
concept of powers of ten
concept of the slide rule as a computing device.

Since I have had to undergo the experience of conversion myself, I am aware of some of the many difficulties which can be encountered in the process, and have tried to produce comments which will help you to overcome these difficulties. One way to make the conversion process simpler is to browse lightly through the earlier chapters of this book once or twice, in order to develop a mental feeling for the whole subject—look at the forest casually before examining the groves and individual trees in great detail.

Good luck!

CHAPTER ONE
What Are SI Units?

SI (for Système International d'Unites) is the international system of units of measurement, identified as SI in all languages. It is the *modernized metric* system—a high refinement of the original metric system first proposed in 1670 and improved upon many times since then.

SI is by far the most superior system of measurement and calculation yet derived. It is logical in concept. It is extremely convenient to use. It provides greater speed in operation. It is the easiest to teach. It is the simplest to learn. Its metric predecessors are already used throughout the scientific world, and are also the measurement and calculation methods used by the great majority of the world's population.

The SI system consists of

1 seven basic units,
2 two supplementary units,
3 sixteen derived units which have special names and any number of coherent interrelated combinations.

In addition to these "official" members of the system, there are certain additional operational techniques which are included in the total system:

4 A structured system of decimal multiples and submultiples expressed as word prefixes,
5 A system of preferred numbers,
6 A collection of general recommendations regarding symbols and abbreviations for units.

Altogether, these official and closely related parts form an integrated system which provides immeasurably simpler measurement and calculation techniques.

In the early chapters of this book, we will investigate the seven topics which comprise the system. Later, we will integrate the parts into applications in a few of the areas of our modern technology.

CHAPTER TWO

Why Use SI Units?

1. FOR CONVERSION'S SAKE

Add: 4 ft $8\frac{1}{2}$ in.
3 ft $6\frac{1}{4}$ in.
5 ft $6\frac{7}{8}$ in.

Solution: First, convert the fractions to equivalent fractions with a common denominator, and add:

$$\tfrac{1}{2} + \tfrac{1}{4} + \tfrac{7}{8} = \frac{4 + 2 + 7}{8} = \tfrac{13}{8}$$

Second, simplify the improper fraction:

$$\tfrac{13}{8} = 1\tfrac{5}{8}$$

Third, rewrite the problem:

Add: 4 ft 8 in.
3 ft 6 in.
5 ft 6 in.
$1\frac{5}{8}$ in.

Fourth, add the inches:

$$8 + 6 + 6 + 1\tfrac{5}{8} = 21\tfrac{5}{8} \text{ in.}$$

Fifth, convert the inches to feet and inches:

$$21\tfrac{5}{8} \text{ in.} = 1 \text{ ft } 9\tfrac{5}{8} \text{ in.}$$

Sixth, rewrite the problem, and add:

$$4 \text{ ft} + 3 \text{ ft} + 5 \text{ ft} + 1 \text{ ft } 9\tfrac{5}{8} \text{ in.} = 13 \text{ ft } 9\tfrac{5}{8} \text{ in.}$$

Now add: 1 m 520 mm
1 m 159 mm
1 m 782 mm

First, rewrite as decimal and add:

$$\begin{array}{r} 1.520\ \text{m} \\ 1.159\ \text{m} \\ 1.782\ \text{m} \\ \hline 4.461\ \text{m} \end{array}$$

Second, adjust to meters and millimeters, if required, which is rare:

$$4.461\ \text{m} = 4\ \text{m}\ \ 461\ \text{mm}$$

Subtract 2 gal $2\frac{1}{2}$ pt from 4 gal $1\frac{1}{4}$ pt

First, borrow 1 pt to rewrite the $1\frac{1}{4}$ pt as $\frac{5}{4}$ pt. Rewrite $\frac{1}{2}$ pt as $\frac{2}{4}$ pt, and subtract:

$$\tfrac{5}{4} - \tfrac{2}{4} = \tfrac{3}{4}$$

Second, borrow 1 gal to rewrite 4 gal 0 pt as 3 gal 8 pt, and subtract:

$$\begin{array}{r} 3\ \text{gal}\ \ 8\ \text{pt} \\ -\ 2\ \text{gal}\ \ 2\ \text{pt} \\ \hline 1\ \text{gal}\ \ 6\ \text{pt} \end{array}$$

Third, simplify 6 pt to 3 qt

Fourth, remember the $\frac{3}{4}$ pt:

$$4\ \text{gal}\ 1\tfrac{1}{4}\ \text{pt} - 2\ \text{gal}\ 2\tfrac{1}{2}\ \text{pt} = 1\ \text{gal}\ 3\ \text{qt}\ \tfrac{3}{4}\ \text{pt}$$

Now subtract: 10.5124 liters from 18.894 liters

$$\begin{array}{r} 18.8940\,\ell \\ -\ 10.5124\,\ell \\ \hline 8.3816\,\ell \end{array}$$

These two examples should illustrate the superior convenience of the metric system. (We have used approximately equivalent metric values only, because it was the *method* we wanted to concentrate on.) The multiples and submultiples of the main units are decimal values, so that there are no common denominators nor arbitrary conversion units to worry about.

Chapter 2

2. FOR COMMUNICATION'S SAKE

When an American newspaperman says "one billion", he means "one thousand million." When an Englishman says "one billion", he means "one million million." If the American had said "giga", everyone would have understood that he was talking about 10^9 as a multiplier. If the Englishman had said "tera", there would have been no doubt that he was using a multiplier of 10^{12}.

The system of decimal multiples and submultiples greatly simplifies expression of very large and very small numbers, while maintaining a universality of communication.

3. FOR EDUCATION'S SAKE

Imagine a classroom where there are no more common denominators and no more arbitrary conversion units; no more fluid ounces, troy ounces, or avoirdupois ounces!

One British authority in 1970 estimated that the monetary value of the simplified teaching of the decimals in the metric system in British schools was at least 20%!

The advance to more useful (and more interesting) topics will be accelerated and appreciated by teachers and students alike. The opportunity to study in greater depth in the same length school day and year will greatly enrich the experiences of students in the lower grades.

Students in the technical field will be able to perform their calculations with greater ease and efficiency. The logical relationships of the units will simplify the approximations and eliminate much drudgery from the classroom.

4. FOR RETRAINING'S SAKE

Old timers in the work force will require some retraining in order to become adequately oriented to the metric system.

Once these workers develop a sense of metric size, and achieve a personal relationship with the new units, the basic decimal operations will provide them with an easy avenue of learning.

They will need conversion tables for only the first few days of their retraining period, and after that will rarely have to mentally convert metric units into their imperial equivalents in order to think sensibly about them.

5. FOR THE SAKE OF INTERNATIONAL TRADE

Did you ever try to get a nut or machine screw for an imported automobile or electrical appliance? When all the nations of the world use the same units of measurement and the same dimensional language, international trade will be simpler for designers, manufacturers, and users. It will be easier to obtain parts and to prepare product specifications. Manufacturing nations will be on a more even footing in world markets.

Over 90% of the world's population now uses one form or another of metric units. Most nations of the world have already established SI standards, or are in the process of doing so.

6. FOR THE SAKE OF MANUFACTURING SIMPLICITY

In North America today, you can buy a wide variety of screw threads: Whitworth, Unified Coarse, Unified Fine, Metric Coarse, Metric Fine, and several others. Under metrication, we will eventually reduce our stocks to two basic threads. (After a few years, of course!)

Similarly, other arbitrary sizes and manufacturers' pets will be reduced to a sensible minimum number of choices. The unnecessarily wide variety in sizes and shapes of all kinds of goods will be brought to a sensible level.

7. FOR THE SAKE OF PRICE COMPARISON

How does a consumer know which package of detergent or cereal to buy? The cartons in the supermarket offer such strange weight-price choices as five pounds of detergent for $1.99, 42 ounces for $1.19, and 20 ounces for 63¢. A popular soap may be purchased in three sizes: four pounds for $1.95, 34 ounces for $1.29, and 1 pound for 65¢. Quickly, now! Which is the most economical of each?

The large economy size is sometimes much more expensive in cents per ounce than the smaller packages. When package sizes and weights come down to some kind of standardization, it will be easier for consumers to determine quickly the price per gram or liter.

8. ARGUMENTS AGAINST METRICATION

There are a few arguments against conversion to the metric system:

1 Cost: It will cost a lot of money to convert from foot rulers and yardsticks to suitable metric rulers and metersticks. Even more to convert all commercial scales from pounds to kilograms. All tape measures

will have to be replaced. All thermometers in hospitals will have to be replaced. Speed limit signs on all our highways will have to be replaced. And so on.

Answer: Yes, it will be expensive. But it is costing our economy just as much in lost export trade to metric nations. A very few years' increased exports will completely offset the country's cost of conversion. And conversion can proceed in ordered steps. Most machines can produce metric sizes, and need be replaced only when they wear out. Many hospitals have already converted to Celsius thermometers and kilogram weights, with very little fuss. On the other hand, housewives making cakes will use teaspoons and cups for a long time to come.

2 Chaos: Factories and storekeepers will be driven frantic trying to keep track of imperial quantities and metric devices during the transition period. Motorists could be confused by driving in kilometers per hour instead of miles per hour. The general public will be terribly confused.

Answer: Not necessarily. Certainly storemen and purchasing agents will have to allow for some overlapping of metric and imperial. But color coding imperial parts and assemblies can reduce confusion in stores and on production floors. Similarly, international traffic signs can co-exist with miles-per-hour signs for a transition period. And suitable public education by the mass media as well as in classes in night schools can effectively limit public confusion. Metrication can be made to work if people want it to.

3 Inertia: The most effective barrier to an easy conversion will be the lack of interest from the public and the work force. How can you force a man to learn an entirely new measuring system if he doesn't want to?

Answer: You can't. But with adequate preparation of the work force, emphasizing the advantages of the metric system, especially the economic advantages, people can be encouraged to want to switch. North American hospitals are already switching to metric measurements with very little difficulty. The United Kingdom made the switch effectively over a ten-year period. We in North America can do as well.

CHAPTER THREE

A Brief History of Measurement

Our records of measurement date back to about 6000 BC, when the civilizations along the Nile and on the Chaldean plains regulated the earliest standards. These early dimensions were initially length units related to man's own dimensions; the digit, the thumb, the handbreadth, the handspan, the forearm, and the foot were all used.

The CUBIT, measured from the point of the elbow to the outstretched tip of the middle finger, became the principal length unit by 4000 BC. It was standardized at what is now about 460 mm. It appears that the pyramids were built 500 cubits on a side, and all measurements were in multiples or submultiples of cubits.

The SPAN, measured between the outstretched tips of thumb and little finger, was approximately half a cubit, or 230 mm.

The PALM, or breadth of four fingers, was about one-sixth of a cubit, or 75 mm.

The THUMBBREADTH, or inch, was an eighteenth of a cubit, or 25 mm. (The thumb was also one-twelfth of a foot.)

The DIGIT, the width of the middle of the middle finger, was one-twenty-fourth of a cubit, or about 20 mm.

The FOOT was about two-thirds of a cubit, 4 palms, 12 thumbs, or 16 digits. It measured about 310 mm.

The MERIDIAN MILE also stretches back to 4000 B.C., and is still used by seamen today. It is 4000 cubits, or 1000 Egyptian fathoms, about 1.85 km.

Later the Greeks and Romans set up their own standard mile as 1000 double-step paces (did we mention the 1.5-m pace?), about 1.5 km.

In 1324, King Edward II of England decreed that the official inch was equal to 3 barley corns taken from the middle of the ear and laid end to end. In 1496, King Henry I established the yard as the distance from the tip of his nose to the end of his thumb when his arm was outstretched.

By 1500, Englishmen had the following set of conversions to work with:

3 barley corns	=	1 inch
12 inches	=	1 foot
3 feet	=	1 yard
9 inches	=	1 span
5 spans	=	1 ell
5 feet	=	1 pace
125 paces	=	1 furlong
$5\frac{1}{2}$ yards	=	1 rod
40 rods	=	1 furlong
8 furlongs	=	1 statute mile
12 furlongs	=	1 league

In addition to length units, there were very early standards of weight and volume. In Genesis 23:16 we read that Abraham "bought a burial field for four hundred shekel-weight of silver, according to the weights current among the merchants." Second Samuel 14:26 mentions the *king's weight*, indicating that there was a standard shekel throughout the land of Israel.

In 1215, Magna Carta established the "London quart" as the specified unit of volume.

Each country introduced its own standards, and there was often no relationship between them. We must mention, however, the 360° circle developed by the Chaldeans in ancient times. (Dr. I. Velikovsky believes that the circle was based on the 360-day year of olden time.)

During more recent times, briefly:

1670 – Gabriel Mouton proposed a decimal measurement system to be based on the earth itself.

1791 – France set up the meter as 10^{-7} part of the length of the meridian through Paris from the equator to the North Pole.

1792– 1799
Delambre surveyed the "Paris meridian" between Barcelona and Dunkirk in order to establish the meridional length. And Picard made his experiments with pendulums.

1799 – The standard meter and the standard kilogram were constructed and stored for reference. Thomas Jefferson was unable to persuade the U.S.A. to use the metric system.

1821 – John Quincy Adams rejected U.S. use of the metric system "until a uniform international measurement system could be worked out."

1827 – Babinet suggested that the wavelength of light might be a more accessible standard of length.

1840 – France outlawed all older units of measurement and made the metric system the only legal one.

1856 – Whitworth in England appealed for standardization of manufacturing methods, so that there would be only ten machine ratings up to 100 horsepower, rather than the thirty then current in England.

1866 – Metric units were made legal in England, equal in status with the older statute units.

1875 – A 17-nation Metric Convention in Paris established the International Bureau of Weights and Measures, and set up the General Conference in Weights and Measures. (C.G.P.M.)

1884 – The United Kingdom signed the Metric Convention.

1889 – The U.K. established the National Physical Laboratory, and officially defined the meter as the distance between two marks on the standard platinum-iridium bar, and the kilogram as the mass of the standard kilogram cylinder.

1895 – A British businessman named Skelton wrote a letter to the *Times* of London complaining about the need to import rolled steel girders, at great cost, in order to meet the unscientific requirements of architects and engineers.

1901 – The U.K. established what is now the British Standards Institution. The U.S.A. established the National Bureau of Standards.

May,
1951 – The Hodgson report of the Board of Trade in Great Britain recommended "establishing the sole use of the metric system for all trade purposes."

1951 – The Japanese Diet made the metric system compulsory.

1954 – The triple point of water was defined to occur at 273.16 K.

July 1,
1959 – Canada, the United States, the United Kingdom, New Zealand, and South Africa formally defined the yard to be 0,9144 meter and the avoirdupois pound to be 0,453 592 237 kilogram.

1960 – The International Practical Temperature Scale (IPTS) was defined in terms of six temperature points. SI units were recommended by the CGPM, ISO, and IEC as length, mass, time, temperature, current, and light intensity.

1962 – Decimal divisions of the inch were generally accepted in North America.

1963 – Some dimensional definitions were made legal:

1 yard = 0.9144 m exactly
1 pound = 0.453 592 37 kg exactly

May 24,
1965 – U.K. Government accepted the metric proposals of an organization which is now known as the Confederation of British Industries, with 1975 as the target date for completion of the conversion to SI units.

March,
1966 – The U.K. established the Standing Joint Committee on Metrication to "encourage" the conversion to metric units.

1967 – The second was defined as 9 192 631 770 cycles of Caesium 133.

1968 – The U.S.A. instructed the Secretary of Commerce to make an appraisal of metric units, and of increased U.S. participation in international systems of standardization.

July 24,
1969 – U.K. Minister of Technology set up an advisory committee to look into the legal requirements involved in adopting SI units as the only legal system of units. Later in 1969, the SI system was adopted for higher education in England, and the Government expressed the hope that the U.K. metrication of industry would be complete by 1975.

January 16,
1970 – Canada issued a White Paper which concluded that "the eventual adoption of the metric system should be an objective of Canadian policy."

June 14,
1971 – Fifty-five Canadians were appointed to the new Standards Council of Canada, responsible for promoting and coordinating voluntary standardization in Canada, and Canadian participation in international standardization organizations. The Council has no enforcement authority.

July,
1971 – U.S. Department of Commerce issued its 12-volume "U.S. Metric Study Interim Report", and an additional report to Congress entitled "A METRIC AMERICA: A decision whose time has come", in which Mr. M. H. Stains, Secretary of Commerce, recommended a 10-year conversion to the International Metric System.

July,
1971 – Mr. S. M. Gossage was appointed Chairman of the Canadian Preparatory Commission for Metric Conversion.

January 13,
1972 – The Honorable Jean-Luc Pepin, Canadian Minister of Industry, Trade, and Commerce, appointed twelve members to Mr. Gossage's Preparatory Commission.

January 20,
1972 – Mr. Gossage reported that it "may well be as much as ten years before the familiar inch/pounds form of measurement gives way entirely to meters and liters." The preliminary phase of the work will take as much as three years.

August 18,
1972 – The U.S. Senate, by a voice vote, *unanimously* approved the metric conversion bill presented earlier in August by the Senate Commerce Committee. The (American) Metric Association expected that this action would stimulate the Science and Astronautics Committee of the House of Representatives to early similar action.

Winter,
1973 – The Bulletin of the Canadian Metric Commission announced target dates for the start of conversion to metric of certain sectors of Canadian industry: highway signs during September 1977; weather forecasts during April and September, 1975; grain sales starting in August, 1976.

October 25,
1973 – The United States House Rules Committee deferred action on House Bill HR11035, despite its large number of co-sponsors, because of concerns voiced by representatives of organized labor and the National Federation of Independent Business. However, nearly one hundred major industrial firms in the U.S.A. have already declared their intention to proceed with internal metric conversion, regardless of government inaction.

CHAPTER FOUR

The Seven Basic SI Units

Length	meter	m
Mass	kilogram	kg
Time	second	s
Electric current	ampere	A
Temperature	kelvin	K
Light intensity	candela	cd
Molecular substance	mole	mol

Length

The standard SI unit of length is the meter (m). By applying decimal prefixes (see Chapter 8), short lengths may be described in millimeters and long lengths in kilometers. Later we will investigate the full range of other multiples and submultiples. Scale drawings are usually made in millimeters, land surveyors use meters, but road maps use kilometers.

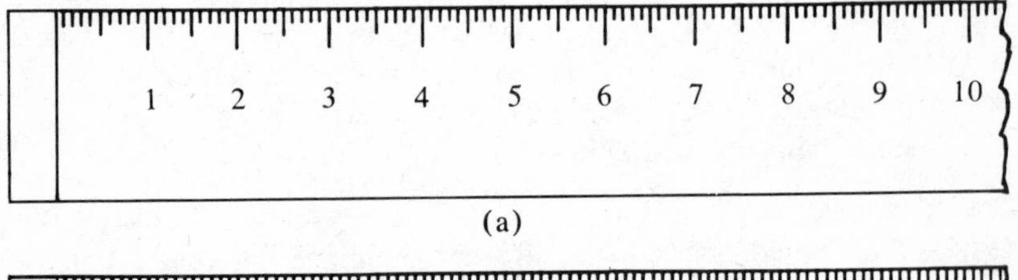

Fig. 4-1 *a*) Metersticks usually show numbers for centimeters.
b) Sometimes metersticks are numbered by millimeters.

Originally, the meter was to have been 1×10^{-7} of the length of the line of longitude passing through Paris from the equator to the North Pole. In 1799, a standard meter was constructed and safely stored so that secondary standards ("substandards") could be made from it for use in other countries.

The present international definition of the standard meter is that it is the length of 1 650 763.73 wavelengths in vacuum of the radiation corresponding to the unperturbed transition between the energy levels $2P_{10}$ and $5d_5$ of the Krypton-86 atom. This orange-red line has a wavelength of $6\,057.802 \times 10^{-10}$ m.

Laboratories, such as that at the National Bureau of Standards in Washington, can reproduce this natural phenomenon whenever it is necessary to do so, in order to check their standards.

Otherwise, other equivalent light wavelengths may be calculated and used.

1 meter = 1.093 61 yards
3.280 839 9 feet
39.370 079 inches

Routine length measurements will be made with metric rulers, metersticks, steel or cloth tapes, etc.

Mass

The standard SI unit of mass is the kilogram (kg). Small masses may be described in grams (g) or milligrams (mg), and large masses as metric tonnes (t), where 1 tonne = 10^3 kg. A liter of pure water at standard temperature and pressure has a mass of 1 kg, within 1 part in 10^4.

The present international definition of the standard kilogram is that it is the mass of a special cylinder of platinum-iridium alloy in the custody of the International Bureau of Weights and Measures which is called the International Prototype Kilogram.

1 kilogram = 2.204 622 6 avoirdupois pounds
35.273 962 avdp ounces

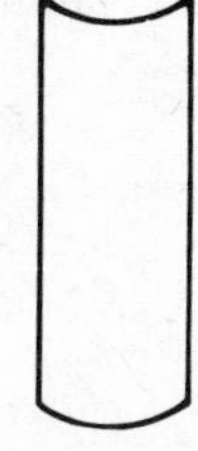

Fig. 4-2 The International Prototype Kilogram is a cylinder of platinum preserved in a storage chamber at Sèvres, France, in the custody of the International Bureau of Weights and Measures.

The mass of a body is revealed by reason of the weight which the gravitational attraction of the earth (or moon, or other "host") gives to that body. Since, for most people, the earth is the only important source of gravitational force, the mass of any body is revealed by its weight on earth. Accordingly, a man describes his mass in terms of his weight on earth. If he goes to the moon his (relatively) unchanged mass will have a different weight because of the moon's weaker gravitational pull.

Normally, then, ordinary people will talk about *weight* as if it were mass. Routine measurements of mass will be made on weigh-scales, using springs or counterbalances.

Time

The standard SI unit of time is the second (s). Despite attempts to use the powers-of-ten multipliers for long periods of time, the minute (min) (= 60 seconds), and the hour (h) (= 3600 seconds) are still the international time units. Other long periods of time are still the day, week, month, and year. Short periods of time may be described in milliseconds and microseconds.

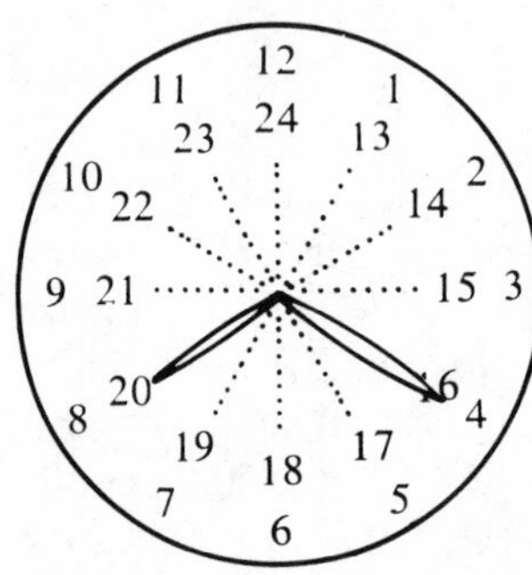

Fig. 4-3 The 24-hour clock is the international standard way of telling time. While most other international units are multiples and submultiples of ten, the hour and minute are still the official international multiples of the second.

The earth has failed to be a satisfactory standard for measuring time, first because of the different ways it provides for time measurement, and second, because the rotational speed of the earth is not absolutely constant.

1 calendar year = 365 mean solar days
1 sidereal year = 365.242 mean solar days
1 leap year = 366 mean solar days
1 calendar year = 12.3600 lunar months
1 tropical year = 365.242 mean solar days
1 lunar month = 29.530 58 mean solar days

The present international definition of the second is that it is the time duration of 9 192 631 770 periods of the radiation of the transition between the two hyperfine levels of the ground state of the atom of Caesium 133.

The second is also defined as $\frac{1}{86\,400}$ of the mean solar day.

The ephemeris second is defined to be exactly $\frac{1}{31\,556\,925.974\,7}$ of the tropical year of 1900, measured at noon on January 1, 1900!

1 mean solar second = $1.157\,407\,4 \times 10^{-5}$ mean solar day
$1.160\,576\,3 \times 10^{-5}$ sidereal day
0.000 277 7 mean solar hours
0.000 278 538 sidereal hours
0.016 66 mean solar minutes
0.016 712 2 sidereal minutes
1.002 737 sidereal seconds

Routine measurements of time will be by watches and clocks. Special radio stations broadcast exact time signals for calibration purposes. The international clock is a 24-hour clock, so that there is no confusion between A.M. and P.M.

Intensity of Electric Current

The standard SI unit of electric current is the ampere (A). Decimal prefixes let us describe milliamperes (mA) and microamperes (μA), and so on. Larger amounts of current might be described in kiloamperes (kA).

A simple definition of the ampere is that it is the intensity of current flow through a 1-ohm resistance under a pressure of 1 volt of potential difference.

The international definition of the ampere is that it is the constant current which, if maintained in two straight parallel conductors of infinite length, of negligible cross-sectional area, and placed exactly 1 meter apart in a vacuum, will produce between them a force of 2×10^{-7} newton per meter length of wire.

1 ampere = 1 coulomb per second
$1.036\,086 \times 10^{-5}$ Faradays
current which will deposit 0.001 182 g of silver in 1 second

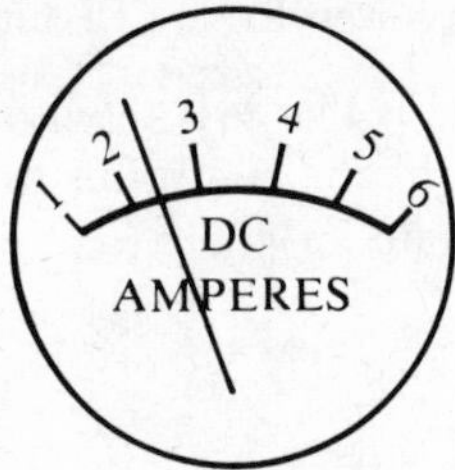

Fig. 4-4 Routine measurement of electric current will be made by ammeters.

Routine measurements of electric current will be made by ammeters such as the analogue meter shown in Figure 4-4, or by digital ammeters, either of which have been compared with high quality meters which have been calibrated by comparison with substandards.

Temperature

The standard SI unit of temperature is the kelvin (K) (*not* degree Kelvin). The freezing point of pure water is 273.15 K.

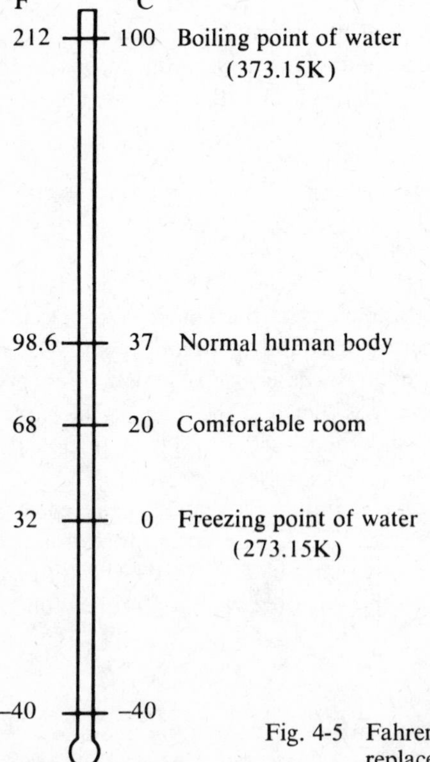

Fig. 4-5 Fahrenheit temperature readings will be replaced by Celsius (formerly Centigrade).

Ordinary temperature measurements are made with the Celsius (formerly centigrade) scale, on which the freezing point of pure water is zero degrees Celsius (0°C). A change of 1 degree Celsius is equal to a change of 1 kelvin, so 0°C = 273.15 K. Normal human body temperature is 37°C.

Scientific temperature measurements above 90 K are made on the International Practical Scale of Temperature (IPST), defined in 1948 by means of the boiling, triple, and freezing points of water, which can be related, in turn, to such devices as platinum resistance thermometers for ease of measurement. Very low temperatures may be measured by a helium gas pressure scale.

The present international definition of the kelvin is that it is $\frac{1}{273.16}$ of the thermodynamic temperature of the triple point of pure water.

Luminous Intensity

The standard SI unit of luminous intensity is the candela (cd). One candela will produce a luminous flux of 1 lumen within a solid angle of 1 steradian.

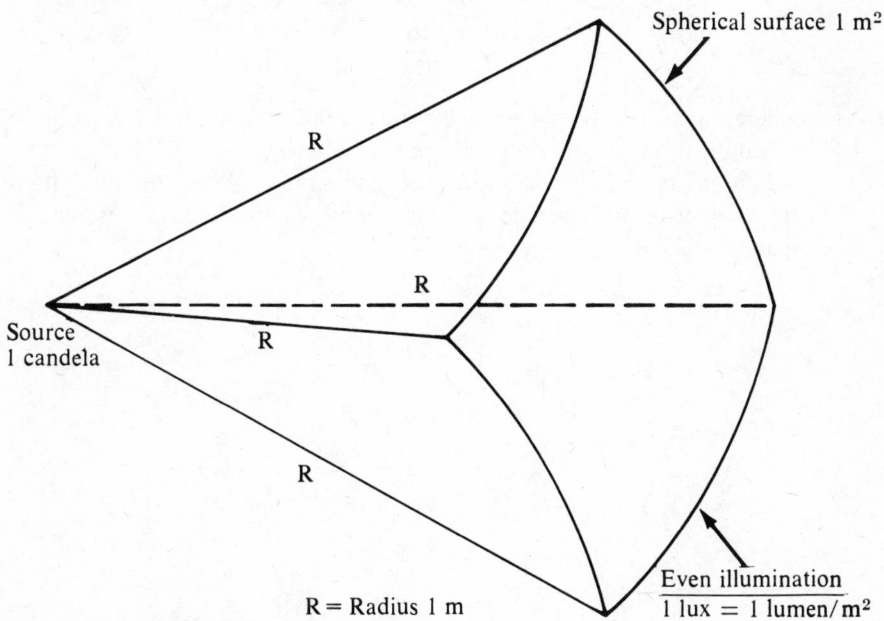

Fig. 4-6 Illumination measurements are based on the amount of radiation falling on a curved spherical surface 1 meter square, 1 meter from the light source.

The older unit of measure was the candle, and you may have seen lightmeters calibrated in footcandles, where 1 footcandle represented the light intensity of 1 lumen per square foot.

The present international definition of the candela is that it is the luminous intensity, perpendicular to the surface, of $\frac{1}{600\,000}$ m^2 of a "black body" at the temperature of freezing platinum under a pressure of 101 325 newtons per square meter (pascals).

Molecular Substance

(This is essentially a unit "for scientists only".)

The standard SI unit for the amount of molecular substance is the mole. One mole of any substance is the *gram molecular weight* of that material. Since a molecule of water consists of two atoms of hydrogen and one of oxygen, a mole of water weighs:

$$\begin{array}{ll} H_2 = 2 \times 1.008 & = 2.016\ \text{g} \\ 0 \ \ = 1 \times 16 & = 16\ \ \text{g} \\ \hline 1\ \text{mol of}\ H_20 & = 18.016\ \text{g} \end{array}$$

where the atomic weight of hydrogen is 1.008, and that of oxygen is 16, based on the *relative* weights of the two substances.

Physicists and chemists will relate the mole to Avogadro's Constant: the number of molecules to make up a mole of any substance is approximately 6.023×10^{23}!

CHAPTER FIVE

Two Supplementary SI Units

Plane angle	radian	rad (. . .^{r})
Solid angle	steradian	sr

Plane Angle

The standard SI unit of plane angle is the radian (rad *or* . . .^{r}), which is the angle formed between two radii of a circle and subtended by an arc whose length is equal to a radius.

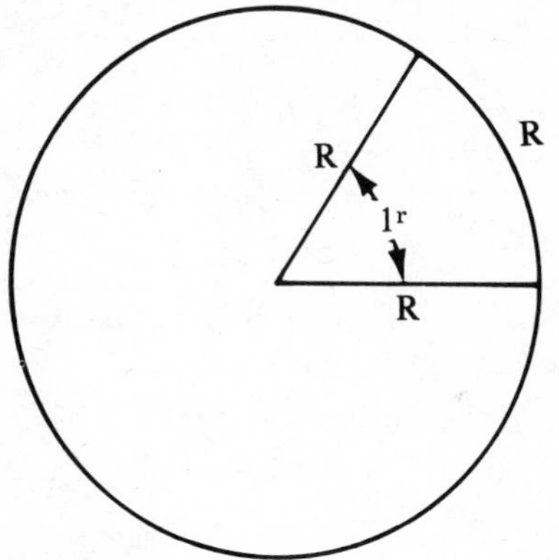

Fig. 5-1 The radian is the angle bounded by two radii, and subtended by an arc of radius length. $2\pi^r = 360°$

$$1 \text{ circle} = 2\pi \text{ radians } (2\pi \text{ rad } or\ 2\pi^r)$$

Ordinary measurements will still be made in degrees (deg, . . .°), where 1 circle = 360°. Although some trades will subdivide the degree into the traditional 60 minutes and 3600 seconds, others will use tenths and hundredths of degrees for the sake of easy slide rule and interpolated calculations.

$$2\pi^r = 360°$$

$$1^r = \frac{180°}{\pi} \simeq 57.29°$$

Chapter 5

Solid Angle

The standard SI unit of solid angle is the steradian (sr), which is the solid angle at the center of a sphere which subtends an area on the spherical surface which is equal to that of a square whose sides are equal to the radius.

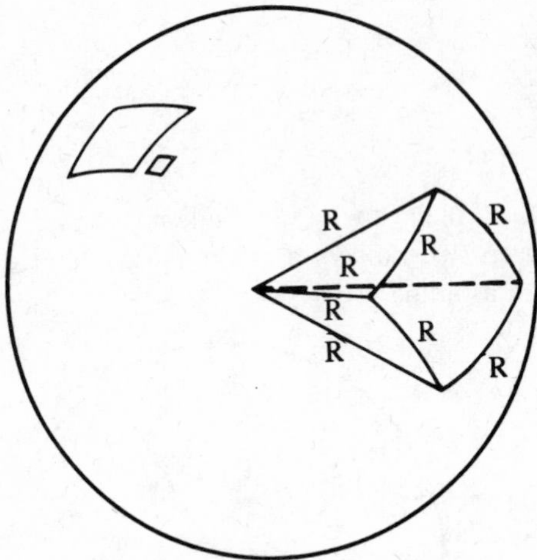

Fig. 5-2 One steradian is the solid angle at the center of a sphere which measures out a surface area whose sides are almost radius lengths.

CHAPTER SIX

Sixteen Derived SI Units Which Have Special Names

All the SI units may be interrelated and combined according to the usual techniques of mathematics and physics. A magnetic flux of 25 webers crossing through an area of 2 square meters constitutes a flux density of 12.5 Wb/m^2. A body with a mass of 50 kg and a volume of 10 m^3 has a density of 5 kg/m^3. It is not really essential that there be a special name for the unit Wb/m^2 or the unit kg/m^3. Electricians do not really require a special name for volts per ohm or for cycles per second.

However, there are many units which are used so often that it is convenient to give them names of their own, and, at present, there are sixteen such units in the SI system.

Energy (work, heat quantity)	joule	J	kg•m^2/s^2	= N•m	
Force	newton	N	kg•m/s^2	= J/m	
Pressure	pascal	Pa	N/m^2		
Power	watt	W	kg•m^2/s^3	= J/s	
Electric charge	coulomb	C	A•s		
Electric potential difference (voltage, electromotive force)	volt	V,E	kg•m^2/s^3•A	= J/A•s	= W/A
Electric resistance	ohm	Ω	kg•m^2/s^3•A^2	= V/A	
Electric conductance	siemens	S			= 1/Ω
Electric capacitance	farad	F	A^2•s^4/kg•m^2	= A•s/V	
Electric inductance	henry	H	kg•m^2/s^{2}A^2	= V•s/A	
Frequency	hertz	Hz	cycle/s	= s^{-1}	
Magnetic flux	weber	Wb	kg•m^2/s^2•A	= V•s	
Magnetic flux density	tesla	T	kg/s^2•A	= V•s/m^2=Wb/m^2	
Luminous flux	lumen	lm	cd•sr		
Illumination	lux	lx	cd•sr/m^2	= lm/m^2	
Customary temperature	degree Celsius	°C	°C	= K − 273.15	

Chapter 6

Energy

The standard SI unit of energy is the joule (J). The mechanical definition of the joule is that it is the work done when a force of 1 newton moves a distance of 1 meter in the direction of its application.

The electrical unit of energy is the kilowatt-hour (kWh), which is equal to 3.6×10^6 J.

In physics, the unit of energy is the electronvolt (eV), which is equal to $(1.602\,10 \pm 0.000\,07) \times 10^{-19}$ J.

1 joule = $1\ \text{kg}\cdot\text{m}^2/\text{s}^2$
$1\ \text{N}\cdot\text{m}$
0.000 948 45 thermochemical BTU
0.737 562 ft-lb

Force

The standard SI unit of force is the newton (N), which is defined as that force which, when applied to a body having a mass of 1 kilogram, gives it an acceleration of 1 meter per second squared. (You will perhaps notice that g, the acceleration due to gravity, does not appear in this definition. Accordingly, whenever a force is applied in opposition to gravity, g must be considered. $g = 9.81\ \text{m}/\text{s}^2$).

Referring to the definition of energy, above, you can see that

1 newton = 1 J/m
$1\ \text{kg}\cdot\text{m}/\text{s}^2$
10^5 dynes
0.224 809 lb force

Pressure

The standard SI unit of pressure is the pascal (Pa). Another common name is newton per square meter (N/m^2). This is the unit which describes force delivered over a given area, such as a thumb pushing a tack into a notice board, or compressed air holding out the walls of an automobile tire.

1 pascal = $1\ \text{N}/\text{m}^2$
$1.450\,38 \times 10^{-4}\ \text{lb}/\text{in.}^2$
1×10^{-5} bars

Power

The standard SI unit of power is the watt (W), which is the power which generates energy at the rate of 1 joule per second.

1 watt = 1 J/s
3.414 42 thermochemical BTU/h
44.253 7 ft-lb/min
0.001 341 02 horsepower

Electric Charge

The standard SI unit of electric charge is the coulomb (C), which is the quantity of electricity moved during one second by a current of one ampere. The coulomb is also defined as $6.241\,96 \times 10^{18}$ electrons.

1 coulomb = 1 A•s
$1.036\,086 \times 10^{-5}$ Faradays
$6.241\,96 \times 10^{18}$ electronic charges

Electric Potential Difference

The standard SI unit of potential difference is the volt (V). Other common names are electromotive force (emf) and merely "voltage".

This is the potential difference which will cause a current flow of 1 ampere between two points in a circuit when the power dissipated between those two points is one watt.

A simpler definition says that a potential difference of 1 volt will drive a current of 1 ampere through a resistance of 1 ohm.

1 volt = 1 W/A
1 J/A•s
1 $kg{\cdot}m^2/s^3A$
1 A•Ω

Electric Resistance

The standard SI unit of resistance is the ohm (Ω). This is the resistance of a passive circuit element which will limit the current flow through itself to one ampere when a potential difference of one volt is applied to it.

1 Ω = 1 V/A
1 $kg{\cdot}m^2/s^3{\cdot}A^2$

Chapter 6

Electric Conductance

The standard SI unit of conductance is the siemens (S). Conductance is the reciprocal of resistance. A passive circuit element which has a conductance of one siemens will allow a current flow of one ampere when a potential difference of one volt is applied to it.

Formerly, the unit of conductance was simply called the *mho* (*ohm* backwards), and some electrical engineers and technicians used an upside-down Ω as a symbol: ℧.

$$
\begin{aligned}
1\ \mathrm{S} &= 1/\Omega \\
& 1\ \mathrm{A/V}
\end{aligned}
$$

Electric Capacitance

The standard SI unit of capacitance is the farad (F). This is the capacitance which exhibits a potential difference of 1 volt when it holds a charge of 1 coulomb.

$$
\begin{aligned}
1\ \mathrm{F} &= 1\ \mathrm{C/V} \\
& 1\ \mathrm{A{\cdot}s/V} \\
& 1\ \mathrm{A^2{\cdot}s^4/kg{\cdot}m^2}
\end{aligned}
$$

Electric Inductance

The standard SI unit of inductance is the henry (H). This is the inductance of a circuit in which an electromotive force of 1 volt is developed by a current change of one ampere per second.

$$
\begin{aligned}
1\ \mathrm{H} &= 1\ \mathrm{V{\cdot}s/A} \\
& 1\ \mathrm{kg{\cdot}m^2/s^2{\cdot}A^2}
\end{aligned}
$$

Frequency

The standard SI unit of frequency is the hertz (Hz), which is a frequency of one cycle per second.

$$1\ \mathrm{Hz} = 1/\mathrm{s}$$

Magnetic Flux

The standard SI unit of magnetic flux is the weber (Wb). This is the amount of flux which produces an electromotive force of one volt in a one-turn conductor as it reduces uniformly to zero in one second.

$$
\begin{aligned}
1\ \mathrm{Wb} &= 1\ \mathrm{V{\cdot}s} \\
& 1\ \mathrm{kg{\cdot}m^2/s^2{\cdot}A} \\
& 1 \times 10^8\ \text{lines of flux}
\end{aligned}
$$

Magnetic Flux Density

The standard SI unit of flux density is the tesla (T), which is a density of 1 weber per square meter.

$$
\begin{aligned}
1\ \mathrm{T} &= 1\ \mathrm{Wb/m^2} \\
&\quad 1\ \mathrm{kg/s^2 \cdot A} \\
&\quad 1\ \mathrm{V \cdot s/m^2}
\end{aligned}
$$

Luminous Flux

The standard SI unit of luminous flux is the lumen (lm). This is the amount of luminous flux emitted by a uniform point source whose intensity is one candela into a solid angle of 1 steradian.

$$
\begin{aligned}
1\ \mathrm{lm} &= 1\ \mathrm{cd \cdot sr/m^2} \\
&\quad 0.079\ 577\ 4\ \text{candlepower}
\end{aligned}
$$

Luminous Flux Density (Illumination)

The standard SI unit of illumination is the lux (lx). This is the density of radiant flux of 1 $\mathrm{lm/m^2}$.

$$
\begin{aligned}
1\ \mathrm{lx} &= 1\ \mathrm{lm/m^2} \\
&\quad 1\ \mathrm{cd \cdot sr/m^2} \\
&\quad 0.092\ 903\ 0\ \text{ft-c} \\
&\quad 0.092\ 903\ 0\ \mathrm{lumens/ft^2} \\
&\quad 10^{-4}\ \text{phots}
\end{aligned}
$$

CHAPTER SEVEN

Thirteen Other Derived SI Units

In addition to the derived units which have names of their own, there are thirteen other derived units which simply carry the units of the original units which are combined. Most readers will be able to obtain sufficient information from the tabulation below to meet their calculation needs.

Area	square meter	m^2
Volume	cubic meter	m^3
Density	kilogram per cubic meter	kg/m^3
Velocity	meter per second	m/s
Acceleration	meter per second squared	m/s^2
Angular velocity	radian per second	rad/s
Angular acceleration	radian per second squared	rad/s^2
Kinematic viscosity	square meter per second	m^2/s
Dynamic viscosity	pascal-second	Pa·s
Electric field strength	volt per meter	V/m
Magnetomotive force	ampere	A
Magnetic field strength	ampere per meter	A/m
Luminance (brightness)	candela per square meter	cd/m^2

CHAPTER EIGHT

Multiples and Submultiples of SI Units

1. THE CONCEPT OF POWERS OF TEN

A convenient shorthand method of indicating multiples and submultiples of ten is the method of powers, where the *index* or *superscript* number indicates the number of times that the *base* number is multiplied by itself. In general,

$$a \times a \times a \times a = a^4$$

Although a here represents any number, we will confine our present efforts to the base number ten. Based on the above definition:

$$10 \times 10 = 10^2$$
$$10^3 = 10 \times 10 \times 10$$

It should be apparent that $10^1 = 10$
and it can be shown that $10^0 = 1$
10^3 may sometimes be written 1×10^3, and so on.

Example 1: Rewrite 100,000 in powers of ten.

Solution: $100{,}000 = 10 \times 10 \times 10 \times 10 \times 10 = 10^5$

Alternate solution: To rewrite 100,000 to $1 \times 10^{\text{something}}$ requires moving the decimal point 5 places to the left. This is offset by making the power of ten 5:

$$100{,}000 = 10^5$$

Example 2: Rewrite 0.0001 in powers of ten.

Solution: 0.0001 requires that the decimal point be moved 4 places to the right. This is offset by making the power of ten -4:

$$0.0001 = 1 \times 10^{-4}$$

Now study Table 8-1 and see how any number may be expressed as a power of ten.

TABLE 8-1

POWERS OF TEN

Number	Power of 10	Pronunciation
1 000 000	10^6	ten to the sixth power
100 000	10^5	ten to the fifth power
10 000	10^4	ten to the fourth power
1 000	10^3	ten to the third power (ten cubed)
100	10^2	ten to the second power (ten squared)
10	10^1	ten to the first power (ten)
1	10^0	ten to the zero power (one)
0.1	10^{-1}	ten to the minus first power (ten to the minus one)
0.01	10^{-2}	ten to the minus two
0.001	10^{-3}	ten to the minus three
0.000 1	10^{-4}	ten to the minus four
0.000 01	10^{-5}	ten to the minus five
0.000 001	10^{-6}	ten to the minus six

Usually, the *preferred* powers of ten are the *third* powers.

1×10^4 would usually be written 10×10^3

1×10^2 would usually be written 0.1×10^3

2. WORD PREFIXES FOR POWERS OF TEN

Most people today have been exposed to at least a few of the word prefixes which represent powers of ten. You have heard of kilocycles to represent one thousand cycles. You may have heard of a microsecond as a millionth of a second. Table 8-2 gives the prefixes and abbreviations used to represent the preferred third power of ten.

TABLE 8-2

POWER-OF-TEN PREFIXES AND ABBREVIATIONS

Power of ten	Number	Prefix	Abbreviation
10^{-18}	0.000 000 000 000 000 001	atto	a
10^{-15}	0.000 000 000 000 001	femto	f
10^{-12}	0.000 000 000 001	pico	p
10^{-9}	0.000 000 001	nano	n
10^{-6}	0.000 001	micro	μ
10^{-3}	0.001	milli	m
10^0	1.000		
10^3	1 000.	kilo	k
10^6	1 000 000.	mega	M
10^9	1 000 000 000.	giga	G
10^{12}	1 000 000 000 000.	tera	T

In addition to these preferred prefixes, there are a few commonly used prefixes which are still used, although their general use is discouraged.

10^2	hecto	h
10^1	deca or deka	da*
10^{-1}	deci	d
10^{-2}	centi	c

You should try to avoid using these prefixes except in very special circumstances, and in the special names which are indicated later in this book.

3. PRONOUNCING THE PREFIXES

When you pronounce the prefixes, always put the first emphasis on the prefix:

kilometer is properly pronounced KIL-O-meter, not kil-AH-meter
micrometer, as a distance, is pronounced MIC-RO-meter. The measuring device is still pronounced micrAH-meter

* A Special Paper published by the Canadian Standards Association in June, 1973, specifies da as the approved abbreviation for deca. Some agencies have recommended dk as a more suitable, less confusing abbreviation.

CHAPTER NINE

Special Notation for SI Units

There are a few recommended practices to be observed in using SI units, and you should become familiar with them.

1. DECIMALS

When writing a decimal number smaller than 1, always precede the decimal point with a zero:

0.023, not just .023

In most European countries, a comma is used as a decimal indicator instead of a period:

1,087 instead of 1.087

In England, the period is used instead of the comma. From 1968 until the spring of 1972, the English decimal point was indicated in two approved ways:

Printed copy inserted the decimal point above the line:

1·087 0·023

Typed copy was "allowed" to place the decimal point on the line:

1.087 0.023

However, in the spring of 1972, the British Standards Institution amended their standard 6031.1968 to always locate the decimal point on the line.

It is possible that, in North America, the decimal marker will be a comma, and in keeping with a majority vote of the members of the Canadian Metric Association, a comma decimal marker will be used in the remaining part of this book.

2. LARGE NUMBERS

Because of the European use of the comma as a decimal marker, the preferred method of setting large numbers in easy-to-read groups of three is to separate the groups with a space:

2,867,411 is written 2 867 411

Numbers consisting of only four digits *may* be written without the space:

4286 *or* 4 286

Small numbers which consist of long strings of digits will also be written with the space:

0.00235689 will be written 0.002 356 89

but 0.002 3 may be written 0.0023

3. SYMBOLS AND ABBREVIATIONS

– Symbols for physical quantities and physical units should be printed in roman (upright) type:

25 millimeters may be abbreviated 25 mm, never 25 *mm*

– Abbreviations for words should be printed in the same roman type:

electromotive force is abbreviated emf, never *emf*

– There are no periods used at the end of abbreviations, or between letters of compound abbreviations:

emf, never e.m.f.

– There is no *s* added for plurals: the symbol or abbreviation is the same for both the singular and plural:

1 mm 10 mm 0.5 mm

– Symbols and abbreviations are printed in lower case letters unless the word is derived from a proper name:

emf, never EMF
12 meters = 12 m
10 volts = 10 V
20 webers = 20 Wb

– Only SI units are abbreviated. All other units are spelled out:

100 km *but* 10 nautical miles

– Units are spelled out if there is any possibility of doubt:

0.1 1 should be spelled out: 0.1 liter or a suitable defined symbol should be adopted. A script letter ℓ is acceptable for **liter**, and 0.1 ℓ is quite acceptable. This is the symbol used throughout the balance of this book. Some organizations use an italic *l*, but this is not standard, and it should not be used.

4. MATHEMATICAL OPERATIONS

– Multiplication may be indicated by uniting the units, with no space, no hyphen, no other symbol:

$$10 \text{ kg} \times 5 \text{ m} = 50 \text{ kgm}$$

A preferred method of indicating multiplication is the raised dot:

$$10 \text{ kg} \times 5 \text{ m} = 50 \text{ kg}\cdot\text{m}$$

– Division may be indicated by three methods:

$$\frac{\text{m}}{\text{s}} \qquad \text{m/s} \qquad \text{m}\cdot\text{s}^{-1}$$

– When more than one division is required, only one fraction bar or oblique stroke is used:

meters per second per second is written

$$\frac{\text{m}}{\text{s}^2} \quad \textit{or} \quad \text{m/s}^2 \quad \textit{or} \quad \text{m}\cdot\text{s}^{-2}$$

5. PREFIXES

Decimal multiples and submultiples are indicated by single prefixes:

22 pF never 22 $\mu\mu$F
12 ns never 12 mμs

6. POWERS ON UNITS

Symbols of units which are composed of decimal prefixes and basic symbols are considered to be new symbols. These new symbols may be raised to positive or negative powers.

mm^3 means $(\text{mm})^3$ or $(0.001 \text{ m})^3$
not 0.001 m^3
nor 0.001^3m

ps^{-1} means $(ps)^{-1}$ or $(10^{-12}s)^{-1}$
not $10^{-12}s^{-1}$
but something per picosecond

Numerical powers are preferable to word prefixes. Write 25 m^2 rather than 25 sq m.

7. EXPRESSION OF VALUES

Third powers of ten are preferred — both positive and negative. For this reason, values need not always be written in the old "scientific" or "engineering" notation as a number between 1 and 10 times the proper power of 10. Where once we insisted that a number be written

$$1.25 \times 10^5$$

we now look for

$$125 \times 10^3$$

or

$$0.125 \times 10^6$$

Values between 0.1 and 1000, with suitable third power decimal prefixes, will become more and more the standard acceptable range of expression.

When approximating prior to performing a sliderule computation, it will probably still be most convenient to work with numbers between 1 and 10 times the appropriate power of 10.

Values will not normally be mixed. We will write

10.8 kg, not 10 kg 800 g

8. NORMAL EXPRESSION OF LENGTH

Although the meter is the standard unit of length, engineering and architectural drawings will usually indicate length in millimeters. A panel on a blueprint marked 1200 by 2400 will normally be understood to be 1200 mm by 2400 mm. This is the standard practice in the U.K., and will probably apply in North America.

CHAPTER TEN

Preferred Number Systems for SI Units

1. WHAT ARE PREFERRED NUMBERS?

If you have ever had anything at all to do with resistors for electronic circuits, you have some idea of what the subject of preferred numbers is all about. It would be impossible to mass-produce every possible value of resistance required, so selected values are produced. Any resistance value you require can be matched with an official value off the shelf within ±20%, or ±10% or ±5%. If you need 255 ohms, you select a 270-ohm resistor. The error is only 15 ohms in 255, so the percentage error is only 5,9%.

An illustration of the advantage of any system of preferred numbers is shown in Table 10-1. The first column of the table lists the usual values available in an ordinary present-day system, say horsepower ratings of motors, or any other sequence of stepped values from which a customer may choose. Because the sequential values are in steps of 5 or 10, there is a great variation in the percentage increase of one value over the preceding value. The metric arrangement based on the R10 preferred number system uses different size steps in order to achieve a nearly uniform percentage difference between successive steps. The metric system uses fewer steps and achieves more meaningful differences.

There are several ordered systems of preferred numbers, all based upon roots of ten. The R10 system shown in Table 10-1 uses rounded-off values of the various tenth roots of ten. The R5 system uses the fifth roots of ten, and so on.

TABLE 10-1
COMPARISON OF TYPICAL ARITHMETIC AND METRIC SIZES

Typical Arithmetic Series	Increment	% Increase	R10 Series	Increment	% Increase
10			10		
	5	50%		2,5	25%
15			12,5		
	5	33,3		3,5	28
20			16		
	5	25		4	25
25			20		
	5	20		5	25
30			25		
	10	33,3		7	28
40			32		
	10	25		8	25
50			40		
	10	20		10	25
60			50		
	10	16,7		13	26
70			63		
	10	14,3		17	27
80			80		
	10	12,5		20	25
90			100		
	10	11,1			
100					

Example 1: Calculate, to three significant figures, the values of the R5 series of preferred numbers.

Solution: Find, in turn, the fifth roots of 10:

$10^{\frac{1}{5}} = \sqrt[5]{10} = 1.585$

$10^{\frac{2}{5}} = \sqrt[5]{10^2} = 2.508$

$10^{\frac{3}{5}} = \sqrt[5]{10^3} = 3.98$

$10^{\frac{4}{5}} = \sqrt[5]{10^4} = 6.30$

$10^{\frac{5}{5}} = 10.00$

Now compare these values with the official rounded-off values of the R5 series presented in Table 10-2.

TABLE 10-2
R5 SERIES OF PREFERRED NUMBERS

1.0
1.6
2.5
4.0
6.3
10.0

Any such series can be adjusted to cover lower and higher values by multiplying by the appropriate power of 10.

2. THE MOST COMMON SETS OF PREFERRED NUMBERS

By choosing the appropriate roots of ten, we are able to fit intermediate values into the R5 series in order to achieve smaller steps of almost uniform percentage difference. Table 10-3 lists the values of the four most common series of preferred numbers: R5, based, as we have seen, on successive fifth roots of 10; R10, based on the tenth roots of 10; R20, based on the twentieth roots of 10; and R40, based on the fortieth roots of 10.

3. PREFERRED NUMBERS SERIES GIVE NEARLY UNIFORM PERCENTAGE STEPS

Referring to Table 10-1, notice that the percentage differences between the values of the R10 series are all very close to 25%. Using these preferred values, a manufacturer can offer a line of related devices with nearly uniform sequential value differences, in order to better meet the requirements of his customers. What are the percentage steps between adjacent values of the R5 system?

TABLE 10-3

SERIES OF PREFERRED NUMBERS

R5	R10	R20	R40
$\sqrt[5]{10^x}$	$\sqrt[10]{10^x}$	$\sqrt[20]{10^x}$	$\sqrt[40]{10^x}$
1,00	1,00	1,00	1,00
			1,06
		1,12	1,12
			1,18
	1,25	1,25	1,25
			1,32
		1,40	1,40
			1,50
1,60	1,60	1,60	1,60
			1,70
		1,80	1,80
			1,90
	2,00	2,00	2,00
			2,12
		2,24	2,24
			2,36
2,50	2,50	2,50	2,50
			2,65
		2,80	2,80
			3,00
	3,15	3,15	3,15
			3,35
		3,55	3,55
			3,75
4,00	4,00	4,00	4,00
			4,25
		4,50	4,50
			4,75
	5,00	5,00	5,00
			5,30
		5,60	5,60
			6,00
6,30	6,30	6,30	6,30
			6,70
		7,10	7,10
			7,50
	8,00	8,00	8,00
			8,50
		9,00	9,00
			9,50
10,00	10,00	10,00	10,00

Example 2: Determine the percentage steps between the preferred values of the R5 system.
Solution:

Value	Difference	% Increase
1,00		
	0,6	60%
1,60		
	0,9	56%
2,50		
	1,5	60%
4,00		
	2,3	57,5%
6,30		
	3,9	61,9%
10,00		

PROBLEMS 10-1

1 Determine the percentage increases between the values of the R20 system of preferred numbers.
2 Determine the percentage increases between the values of the R40 system of preferred numbers.

4. PREFERRED NUMBER SERIES GIVE NEARLY UNIFORM MAXIMUM TOLERANCE VALUES

In Section 3 of this chapter we saw that the steps of the preferred number series give nearly uniform percentage steps. These steps allow customers to choose the nearest most suitable value from a range of values of devices.

In addition to this convenience, there is another advantage to the preferred number system. This is the advantage of maximum tolerance values, mentioned in Section 1 of this chapter. No matter what value of resistance (or other parameter) you design, there is a standard preferred value which will be close — within a specified tolerance.

Example 3: In the R10 system, two adjacent preferred values are 20 and 25. If you need any value between these preferred values, what is the maximum percentage error introduced by using either 20 or 25?

Solution: The greatest error which will be introduced will occur if the desired value is 22,5 — midway between the two preferred values.

Using 20 instead of 22,5, the error is (−)2,5.

The percentage error is

$$\frac{-2,5}{22,5} \times 100\% = -11,1\%$$

Using 25 instead of 22,5 introduces a percentage error of +11,1%

If any value less than 22,5 is desired, the percentage error will be less than 11,1% when 20 is selected. If any value greater than 22,5 is desired, the error will be less than 11,1% when 25 is selected. The maximum percentage error introduced by using either 20 or 25 in place of any intermediate value is, then, ±11,1%.

PROBLEMS 10-2

1 Choosing values halfway between the preferred values of the R5 system of preferred values, what is the greatest percentage error introduced anywhere in the series?
2 Ditto, R10.
3 Ditto, R20.
4 Ditto, R40.
5 Calculate the three-significant figure values of the R6 series of preferred numbers. What is the nearly uniform percentage increase between steps of the R6 series? What is the maximum percentage error introduced by using the preferred values of the R6 series?
6 Ditto, R12
7 Ditto, R24

5. THE FOUR SIZE CHOICES OF THE CONSTRUCTION INDUSTRY

When an engineer or architect designs a structure, or a sub-assembly, or a single structural element, he will try to assign dimensions according to the following criteria:

1 There will be a large number of products for which dimensional coordination is essential. Included in this "set" of products are door-sets, partitions, and wall panels.

2 There will be a very large number of products which must relate dimensionally to the products of Group 1 above, but need not be absolutely coordinated with them. Bricks and tiles would be included here.

3 There will be many products which do not need to be dimensionally related to products of either Group 1 or Group 2, but which have to fit together among themselves, such as plumbing pipes and all piping fixtures, and electrical conduit.

4 There are many products which need only to have reasonable metric sizes, such as furniture.

5 There will be a great need for measuring devices for use in design, manufacturing, installation, and servicing.

The construction industry in the U.K. has established four choices of sizes to meet the first four of these criteria.

First preference is given to 300 mm, and simple whole number multiples of 300.

Second preference is given to 100 mm, and multiples of 100.

Third choice is simple multiples of 50 mm.

Fourth choice is simple multiples of 25 mm.

The common size of a building brick could then be 300 mm long, by 100 mm high by 50 mm wide.

It would not normally be 280 mm × 110 mm × 65 mm.

6. THE THREE CHOICE RANGES OF ORDINARY VALUES, BASED ON THE PREFERRED NUMBER SYSTEM

A synthesis of the preferred number systems has been prepared in the United Kingdom to give technicians and mechanics some guidance in choosing dimensional and rating values. The complete set comprises three choice ranges, as illustrated in Table 10-4.

TABLE 10-4

FIRST, SECOND, AND THIRD CHOICES OF NUMERICAL VALUES

First choice	Second choice	Third choice	First choice	Second choice	Third choice	First choice	Second choice	Third choice	First choice	Second choice	Third choice	First choice	Second choice	Third choice
1	1	1	6	6	6		28	28			74			138
	1,1	1,1			6,2	30	30	30	75	75	75	140	140	140
1,2	1,2	1,2		6,5	6,5		32	32			76			142
		1,3			6,8			34		78	78		145	145
	1,4	1,4		7	7	35	35	35	80	80	80			148
1,5	1,5	1,5			7,5			36			82	150	150	150
	1,6	1,6	8	8	8		38	38		85	85			152
		1,7			8,5	40	40	40			88		155	155
	1,8	1,8		9	9		42	42	90	90	90			158
2	2	2			9,5			44			92	160	160	160
		2,1	10	10	10	45	45	45		95	95			162
	2,2	2,2		11	11			46			98		165	165
		2,4	12	12	12		48	48	100	100	100			168
2,5	2,5	2,5			13	50	50	50			102	170	170	170
		2,6		14	14		52	52		105	105			172
	2,8	2,8		15	15			54			108		175	175
3	3	3	16	16	16	55	55	55	110	110	110			178
		3,2			17			56			112	180	180	180
	3,5	3,5		18	18		58	58		115	115			182
		3,8			19	60	60	60			118	190	190	190
4	4	4	20	20	20		62	62	120	120	120			192
		4,2			21			64			122		195	195
	4,5	4,5		22	22	65	65	65		125	125			198
		4,8			23			66			128	200	200	200
5	5	5			24		68	68	130	130	130			
		5,2	25	25	25	70	70	70			132			
	5,5	5,5			26		72	72		135	135			

CHAPTER ELEVEN

Everyday Associations in SI Units

In order that SI units may be mastered quickly, it is essential that teachers and tradesmen, as well as students, housewives, and engineers, develop some everyday associations with SI units. People should associate familiar objects with convenient metric equivalents, in order to achieve a working familiarity with SI units.

You should develop such familiarity for yourself as quickly as possible, so that you need not continually refer to tables of equivalents. You must stop thinking in terms of miles per hour, inches of handspan, and quarts of milk. Think at every opportunity in metric quantities in order to quickly achieve an effective mastery of metric units.

MENTAL APPRECIATION OF MAGNITUDE IN METRIC UNITS

Length quantity	*Metric size*	*Significant increment*
Thumbbreadth	20-25 mm	1 mm
Handspan	20-25 cm	1 cm
Height of average man	170 cm	1 cm
Height of normal door	2000 mm	10 mm
Length of average car	4700 mm	100 mm
Size of average room	4.3 m × 3.7 m	0,1 m

You should prepare for yourself, from your own investigation, similar working magnitudes for area, volume, and weight.

CHAPTER TWELVE

SI Conversions — for Oldtimers Only

The poorest way to learn to use the SI system is to make conversions from SI to imperial or vice versa. But many people whose total experience has been with imperial units feel they need conversion tables in order to ease the psychological shock of switching to metric.

We encourage such people to lean more lightly than they think possible upon conversions, and to emphasize the suggestions in Chapter 11. You should make a special effort to quickly develop mental metric associations, and to put imperial units out of your mind. Effective learning of any subject requires practice. Force yourself to think and talk metric. Use the tables only as aids to developing the metric associations.

Do not, under any circumstances, try to change old English maxims into such ridiculous statements as:

28,349 grams of prevention is worth 0,453 kilograms of cure.

Develop your new ideas along the lines of

give him a millimeter, and he'll take a kilometer.

CHAPTER THIRTEEN
Everyday SI Units

While there are recognized standards of measurement for all the usual dimensions, there are many multiples and submultiples used in everyday work. In addition to these official variations of the main standard unit, there are often some other units so widely used that they are still tolerated, even if they do not conform to the standard. In this chapter we will investigate the various everyday units of the main quantities.

1. LENGTH

TABLE 13-1
EVERYDAY LENGTH UNITS

Unit of length	*Symbol*	*Relationship to Standard*
METER	m	$m \times 1$
centimeter	cm	$m \times 10^{-2}$
millimeter	mm	$m \times 10^{-3}$
micrometer	μm	$m \times 10^{-6}$
kilometer	km	$m \times 10^{3}$
International nautical mile	n mile	$m \times 1852$

Notice here the retention of the nautical mile as a length unit. This distance is in such common use by navigators of all nations that it will continue to be used for many years. And while millimeters will be the standard measure for scientists and tradesmen, ladies will reasonably refer to themselves as 91-61-91, thinking in centimeters.

2. AREA

TABLE 13-2
EVERYDAY AREA UNITS

Unit of area	*Symbol*	*Relationship to Standard*
SQUARE METER	m^2	$m^2 \times 1$
square millimeter	mm^2	$m^2 \times 10^{-6}$
are	a	$m^2 \times 10^{2}$
hectare	ha	$m^2 \times 10^{4}$

Notice here the use of second powers of ten for units of area. In practice, thousands of square meters would represent areas too large for convenience.

3. VOLUME

TABLE 13-3
EVERYDAY VOLUME UNITS

Unit of volume	*Symbol*	*Relationship to Standard*
CUBIC METER	m^3	$m^3 \times 1$
cubic millimeter	mm^3	$m^3 \times 10^{-9}$
liter	ℓ or liter	$m^3 \times 10^{-3}$
milliliter	$m\ell$	$\ell \times 10^{-3} = m^3 \times 10^{-6}$

The liter was originally defined to be equal in size to the volume of a kg of pure water at 4°C, and was approximately equal to 10^{-3} cubic meters. In 1964, the twelfth General Conference on Weights and Measures redefined the liter to be equal to 10^{-3} m^3.

4. MASS

TABLE 13-4
EVERYDAY MASS UNITS

Unit of mass	*Symbol*	*Relationship to Standard*
KILOGRAM	kg	$kg \times 1$
gram	g	$kg \times 10^{-3}$
metric tonne	t	$kg \times 10^3 = Mg$

The standard of mass is the kilogram, *not* the gram. You must watch decimal multiples of the kg, and adjust the powers-of-ten prefixes accordingly. One thousand kilograms must be adjusted to

$10^3 \times 10^3$ grams = 1 Mg.   Do not use kkg.

5. FORCE

TABLE 13-5
EVERYDAY FORCE UNITS

Unit of force	*Symbol*	*Relationship to Standard*
NEWTON	N	$kgm/s^2 = N \times 1$
kilonewton	kN	$N \times 10^3$
meganewton	MN	$N \times 10^6$

The newton is a very small force, and you will often find it more convenient to report in kilonewtons as units more likely to appear in normal operations. It has been suggested that an apple tossed into the air strikes the hand, on falling, with a force of about one newton.

6. PRESSURE

TABLE 13-6

EVERYDAY PRESSURE UNITS

Unit of pressure	*Symbol*	*Relationship to Standard*
PASCAL	Pa	Pa × 1
kilonewton per square meter	kPa	Pa × 10^3
bar	bar	Pa × 10^5
millibar	mbar	Pa × 10^2

The bar and millibar are in such current use among meteorologists that they will undoubtedly remain common currency for many years. The pascal is so small that such measurements as tire pressures might well be given in bars instead.

Normal atmospheric pressure of 14.6 **lb/in^2** is equal to 100 663,41 Pa = 0.1 MPa. Tire pressures could well be described in terms of decimal parts of megapascals or hundreds of kilopascals.

7. TIME

TABLE 13-7

EVERYDAY TIME UNITS

Unit of time	*Symbol*	*Relationship to Standard*
SECOND	s	s × 1
millisecond	ms	s × 10^{-3}
microsecond	μs	s × 10^{-6}
nanosecond	ns	s × 10^{-9}
picosecond	ps	s × 10^{-12}
minute	min	s × 60
hour	h	s × 3600
day, month, year, century		

Many scientific measurements in time require very small quantities, and the usual scientific submultiples apply. However, the commoner units of the multiples of seconds will remain in use for many, many years.

In addition to time in hours and minutes, some scientific calculations accept calculations in decimal parts of hours: 4.6h will often be quite as acceptable as 4 h 36 min.

8. SPEED

TABLE 13-8
EVERYDAY SPEED UNITS

Unit of speed	*Symbol*	*Relationship to Standard*
METER PER SECOND	m/s	m/s × 1
kilometer per hour	km/h	$m/s \times \frac{1}{3.6} = m/s \times 0.278$
knot (International nautical mile per hour)	kn	m/s × 0.514 444

9. ENERGY

TABLE 13-9
EVERYDAY ENERGY UNITS

Unit of energy	*Symbol*	*Relationship to Standard*
JOULE	J	$N \cdot m = kg \cdot m^2/s^2$
kilowatthour	kW•h	$J \times 3.6 \times 10^6$

10. POWER

TABLE 13-10
EVERYDAY POWER UNITS

Unit of power	*Symbol*	*Relationship to Standard*
JOULE PER SECOND	J/s	J/s × 1
watt	W	J/s × 1
kilowatt	kW	$J/s \times 10^3$

11. TEMPERATURE

TABLE 13-11
EVERYDAY UNITS OF TEMPERATURE

Unit of temperature	*Symbol*	*Relationship to Standard*
KELVIN	K	K × 1
degree Celsius	°C	K − 273.15
Temperature interval:		
Celsius degree	deg C	1 deg C = 1 K

12. ANGLE

TABLE 13-12

EVERYDAY ANGLE UNITS

Unit of angle	*Symbol*	*Relationship to Standard*
RADIAN	rad $(\ldots^{r})$	rad × 1
degree	$\ldots^{\circ}$	rad $\times \frac{180}{\pi} \simeq$ rad × 57

13. ELECTRICAL UNITS

TABLE 13-13

EVERYDAY ELECTRICAL UNITS

Unit of electrical measurement		*Symbol*	*Relationship to Standard*
Potential difference:	volt	V	W/A
Electric current:	ampere	A	A × 1
Resistance:	ohm	Ω	V/A
Conductance:	siemens	S	A/V
Frequency:	hertz	Hz	1/s
Rotational frequency:	revolution per second	rev/s	1/s
Power:	watt	W	J/s

14. SOME ADDITIONAL TOLERATED UNITS

TABLE 13-14

ADDITIONAL EVERYDAY UNITS

Quantity	*Unit*	*Symbol*	*Relationship to Standard*
length	parsec	pc	$m \times 30.87 \times 10^{15}$
area	barn	b	$m^2 \times 10^{-28}$
	hectare	ha	$m^2 \times 10^4$ ($\simeq$2.47 acres)
volume	liter	l, liter	$m^3 \times 10^{-3}$
pressure	bar	bar	$N/m^2 \times 10^5$
kinematic viscosity	stokes	St	$m^2/s \times 10^{-4}$
dynamic viscosity	pascal-second	Pa•s	$P \times 10^{-1}$
radioactivity	curie	Ci	$37 \times 10^9/s$
energy	electronvolt	eV	$1.602\,1 \times 10^{-19}$ J
	kilowatthour	kW•h	3.6×10^6 J

CHAPTER FOURTEEN
Decimal Calculations for SI Units

14-1 DECIMAL SUBMULTIPLES

The tradesman who has worked all his life in fractions will still recognize basic decimal operations because of our use of decimal currency. Everyone exposed to this book will recognize that \$5,95 represent five dollars and an additional ninety-five hundredths of a dollar. With this ready background you should skim through the following examples and answer the problems in this chapter.

Example 1: 25,261 m represents 25 whole meters and an additional $\frac{261}{1000}$ of a meter, or an additional 261 mm.

Example 2: 12 m + 16 mm would be represented by 12,016 m.

PROBLEMS 14-1

Write the following expressions in terms of their SI defined units and third-power of ten submultiples:

1 124,27 m
2 18,015 m
3 6,193 kg
4 0,535ℓ
5 0,007 m

Write the following expressions as decimal numbers, using only SI units:

6 16 m 244 mm
7 10 kg 100 g
8 22 g
9 515 mℓ
10 25,4 cm

14-2 DECIMAL ADDITION AND SUBTRACTION

Adding and subtracting decimal quantities involves listing the numbers to be operated on, keeping the columns aligned on the decimal points.

Example 3: Add: 12,92 m + 125 mm + 12 mm + 1 m 5 mm

Solution: rewrite each quantity in meters in decimal form and add:

12,92 m	=	12,920 m
125 mm	=	0,125
12 mm	=	0,012
1 m 5 mm	=	1,005
		14,062 m

Example 4: Subtract 485 g from 1 kg 208 g

Solution:

$$\begin{array}{r} 1{,}208 \text{ kg} \\ -\ 0{,}485 \phantom{\text{ kg}} \\ \hline 0{,}723 \text{ kg} \end{array} = 723 \text{ g}$$

PROBLEMS 14-2

Add:

1 12 m, 0,860 m, and 1,003 m
2 5 m, 1,010 m, and 0,098 m
3 120,8 m, 497 mm, and 6 m, 200 mm
4 18,3 kg, 0,65 kg, 440 g, and 5 g
5 2,05ℓ, 1978 mℓ, 80 mℓ, and 135 mℓ

Subtract:

6 4,98 m from 19,807 m
7 12,087 ℓ from 87,65 ℓ
8 47 mℓ from 820 mℓ
9 221 g from 450 g
10 6,241 mm from 4,275 m

14-3 DECIMAL MULTIPLICATION

When two or more decimal numbers are multiplied together, the number of decimal places in the answer is the *sum* of the decimal places in the contributing numbers.

Example 5: Multiply 25,8 by 6,23

Solution:

```
   25,8     1 decimal place
×   6,23    2 decimal places
--------
    774
   516
 1548
--------
 160,734    3 decimal places
```

PROBLEMS 14-3

Multiply:

1 12,9 by 2,87
2 0,0076 by 1,93
3 1,024 m by 0,005 m
4 120,96 m by 2,188 m
5 0,877 kg by 2;006 m

14-4 DECIMAL DIVISION

When dividing by a decimal quantity, multiply both divisor and dividend by the same power of ten in order to convert the divisor to a whole number. Then divide in the usual fashion.

Example 6: Divide 145,228 6 by 2,51

Solution: Multiply both parts by 100, and set out the exercise for long division, locating the decimal point for the answer:

$$251\overline{)14\ 522,86}$$

Divide to obtain 57,86

If the quotient is a "continuing" number, it is taken to some suitable number of decimal places. In the problems which follow, calculate to four places of decimals, and round off to three places.

PROBLEMS 14-4

Divide:

1 4,301 by 2,3
2 96,226 7 by 4,38
3 229,379 2 by 12,1
4 0,121 69 m^2 by 0,28 m
5 0,316 72 J by 0,097 s
6 765,344 02 kg•s by 9,08 s

CHAPTER FIFTEEN

Length Calculations in SI Units

Length Units meter
millimeter
micrometer
kilometer

Draughtsmen, building tradesmen, motorists, dressmakers — almost everyone is concerned with measurements of length. Use your previous knowledge of the fundamentals of length measurements to answer the following problems.

PROBLEMS 15-1 Length measurements
Measure the forty indicated distances on Figures 15-1 and 15-2.

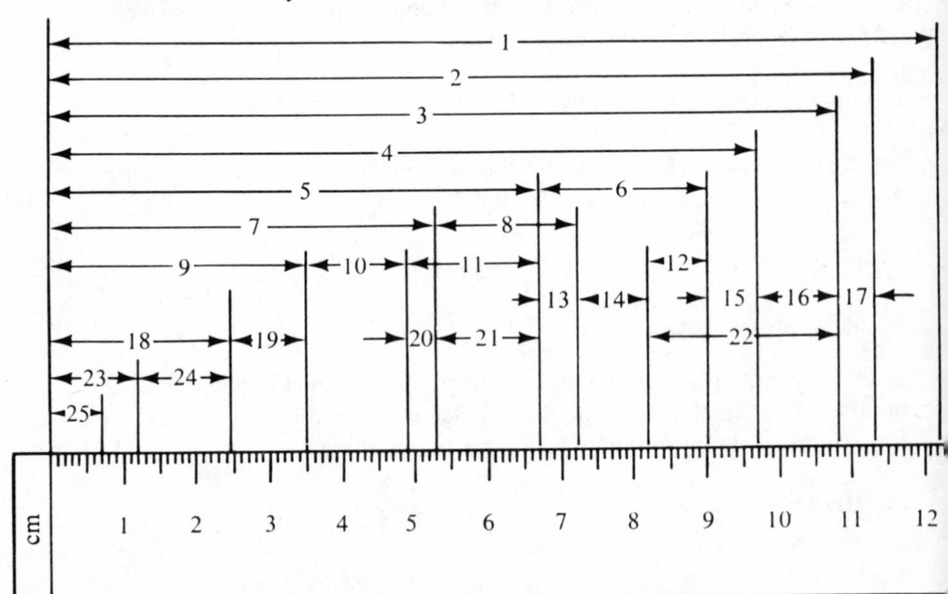

Fig. 15-1 Measure the 25 indicated dimensions in millimeters.

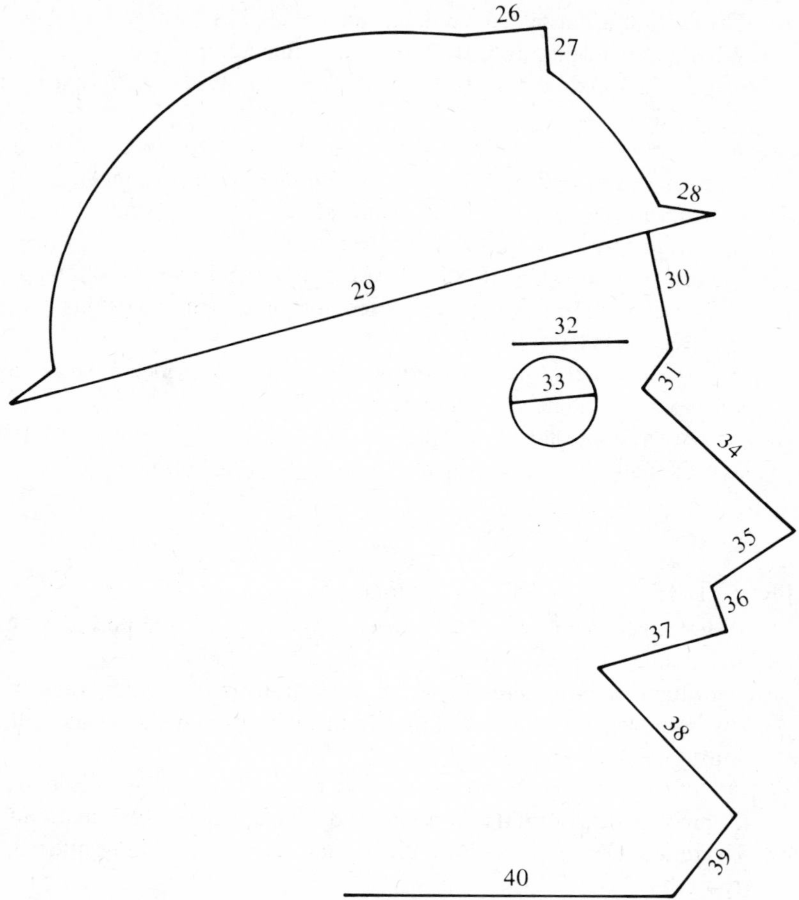

Fig. 15-2 Measure the 15 indicated dimensions in millimeters.

PROBLEMS 15-2 Personal Measurements

1 A man standing at attention measures about $17\frac{3}{4}$ in. across the upper arms. What is this measurement in millimeters (in round numbers)?

2 The same man stands about 5 ft 7 in. tall. What height is this in millimeters (in round numbers)?

3 A woman standing with arms outstretched can just touch her fingertips to two walls 5 ft $4\frac{1}{2}$ in. apart. What is this span, approximately, in millimeters?

4 The eye-level of a seated woman is about 3 ft $7\frac{3}{4}$ in. above floor level. What is this height, approximately, in millimeters?

5 A normal walking pace for a woman is 2 ft. What is this pace in millimeters?

6 A workman kneeling on one knee requires a space of about 3 ft $7\frac{1}{2}$ in. What is this horizontal clearance in millimeters, approximately?

7 A handrail on a staircase is 840 mm above the stair treads. What is this height, rounded off to the nearest inch?

8 The greatest allowable density of standing spectators is 6 persons per square meter. How much floor space in square inches does this allow the average spectator?

9 A man on crutches needs a doorway at least 33 in. wide. What is this width, approximately, in millimeters?

10 A beauty contestant's "statistics" are "35–24–36". What are the approximate metric equivalents of these inch dimensions?

PROBLEMS 15-3 Building Measurements

1 A hotel bedroom measures 3 m by 3,9 m. What is the perimeter of the room?

2 A draughtsman requires 1500 mm for a drawing board, 1500 mm for his reference table, and 900 mm for a half-aisle. What room width must be allowed per draughtsman?

3 An office is to accommodate 4 rows of 6 desks each. Each desk is 750 mm wide and 900 mm is allowed between desks for chairs and clearance. One desk is allowed to touch a wall. How long must the room be?

4 Four designers' cubicles are to be constructed side by side. Each cubicle is to be 1490 mm wide, and each wall involved, including outside walls, is to be 8 mm thick. What total width must be allowed for?

5 Six study carrels are to be constructed side by side. Each carrel will be 990 mm wide. Walls between carrels are to be 10 mm thick, and the walls at the ends of the section are to be 5 mm thick. What total distance must be allowed for?

6 One of the standard sizes of concrete blocks is 200 mm high by 400 mm long by 100 mm thick. The joints between blocks measure 10 mm. Starting from the bottom edge of the bottom block to the top edge of the top block, what is the height of a wall 10 blocks high?

7 How many concrete blocks long is a wall 90 m long?

8 A standard gypsum wallboard is 1200 mm wide by 3000 mm long. Can it be used to provide a single 4-ft by 8-ft panel to repair an existing building?

9 The mandatory height from finished floor to finished floor of a low-rent housing unit is 2.6 m. At an allowable rise for each step of 200 mm, how many steps are required between floors?

10 Rearrange the following sets of lumber cross-sections in order of decreasing area:

a) 2 in. by 4 in.

b) $1\frac{5}{8}$ in. by $3\frac{5}{8}$ in.

c) 50 mm by 100 mm

PROBLEMS 15-4 Traffic Requirements

1 A Jaguar Mark IX automobile is about 16 ft 5 in. long. What round figure describes this length in millimeters?

2 The width of a four-lane dual highway is 14.6 m. What is the equivalent measurement in feet and inches?

3 A sidewalk 1800 mm wide will accommodate two wheelchairs side by side. What is this width in inches, to the nearest $\frac{1}{16}$ in.?

4 In Problem 3, what round number of inches is equivalent?

5 In Problem 3, what round number of feet is equivalent?

6 What round number of kilometers per hour could replace a traffic sign allowing a maximum speed of 30 mi/h?

7 What round number of kilometers per hour could replace 50 mi/h?

8 A roadsign allows a maximum speed of 60 km/h. What round figure in miles per hour is equivalent?

9 A parking lot can accommodate 32 cars side by side. Based on an average width allowance of 2.3 m over open doors, what length of lot is required?

10 Poles of power lines are set 40 m apart. How long will it take to pass between two successive poles at a speed of 50 km/h?

PROBLEMS 15-5 Miscellaneous Length Calculations

1 What is the perimeter of a room 8,3 m long by 6,2 m wide?

2 What is the length of the diagonal across the floor of the room in Problem 1?

3 What is the circumference of a circle whose diameter is 6,25 mm?

4 How far will a car go in 8 min at an average speed of 65 km/h?

5 A motorcycle accelerates from a standstill to 80 km/h in 6 s. What is its acceleration in m/s^2?

CHAPTER SIXTEEN

Area Calculations in SI Units

Area Units square meters
square millimeters
square kilometers
ares
hectares

Measurement of land in acres and sections will probably continue in North America for some time to come. But it will be helpful to begin to become familiar with metric area measurement.

PROBLEMS

1 What is the floor area of a room 12 m by 15 m?

2 If the ceiling height of the room in Problem 1 is 2,5 m, what is the total wall surface?

3 What is the wall area in Problem 2 in ares?

4 A building lot is 50 ft wide and 120 ft deep. Describe its area in
a) square feet
b) acres
c) square meters
d) ares.

5 A section of land is defined variously as 1 mile square, 640 acres, or $\frac{1}{36}$ of a township. What is the area of a quarter-section in
a) square feet
b) acres
c) square meters
d) square kilometers
e) ares
f) hectares?

6 What is the area of a triangle of base 85 mm and altitude 40 mm?

7 What is the area of a circle whose diameter is 25,4 mm?

8 What is the curved surface area of a cylinder 3,5 m high and 750 mm in diameter?

9 What is the surface area of a sphere 50 mm in diameter?

10 Many engineering measurements have traditionally been made in mils, where 1 mil = $\frac{1}{1000}$ in. What is the equivalent square millimeter area of a 400 mil^2 area?

CHAPTER SEVENTEEN
Volume Calculations in SI Units

Volume Units	cubic meters
	cubic millimeters
	liters
	kiloliters

By definition, 1 liter = 1×10^{-3} m^3. Here are some assorted exercises involving volume calculation and conversion from cubic millimeters to liters. In addition, for homemakers, we have included a couple of simple conversions from measuring cup units.

PROBLEMS

1 An imperial quart is often defined as being equal to $1{,}137 \times 10^3$ cm^3. What is this volume in
   a) cubic meters
   b) cubic millimeters
   c) liters?

2 What is the volume of a cube measuring 3,5 mm on a side?

3 What is the volume enclosed by a room 12 m by 15 m by 2,5 m high?

4 How many cubic millimeters are there in 1 cubic meter?

5 What is the volume of a carton 1,6 m long, 800 mm high and 650 mm deep?

6 What is the volume of the carton in Problem 5 in liters?

7 A container is 50 mm in diameter and 110 mm high. What is its volume in
   a) cubic millimeters
   b) cubic meters
   c) liters
   d) cubic inches?

8 A drum is 500 mm in diameter and 1200 mm high. What is its volume in
   a) cubic millimeters
   b) cubic meters
   c) liters?

9 An American bushel is defined to contain 1,244 456 ft^3. How many liters is this?

10 An imperial gallon is defined to contain 277,4 $in.^3$. How many liters is this?

11 The standard measuring cup used by Canadian homemakers contains one-fifth of an imperial quart, ie, 8 fluid ounces. 1 fl oz = 1,733 87 $in.^3$. What is the milliliter volume of a cup?

12 A U.S. pint contains 16 fluid ounces, and the U.S. fluid ounce is 1,804 7 $in.^3$. What is the milliliter volume of a U.S. pint?

13 In the forest industry, a volume of 100 ft^3 of wood products is called a "cunit". What is this volume in cubic meters? What would be a reasonable "metric equivalent cunit"?

CHAPTER EIGHTEEN

Weight (Mass) Calculations in SI Units

Mass Units grams
kilograms
metric tons

The mass of any body is revealed by its weight as a result of the gravitational attraction of its surroundings. The same mass has a different weight on different planets. Since most of the people who use this book will deal only with weight on the earth, we can accept the general notion that "mass" and "weight" mean the same thing to most people. (See Chapter 22 for a differentiated view of mass and weight.)

PROBLEMS

1 The average brick of butter sold in supermarkets weighs one pound. What is the pound-weight of a kilogram brick of butter?
2 A particular bakery's loaf of bread weighs 14 oz. What is this weight in grams?
3 The contents of a bottle of ketchup weighs 252 g. What is this weight in ounces?
4 A standard "short" ton weighs 2000 lb. What is the pound-weight of a metric tonne?
5 Precious metals are weighed in troy ounces instead of avoirdupois. What is the gram equivalent of 20 troy ounces?
6 The density of cast iron is 7.2 Mg/m^3. What is the weight of a billet of cast iron which has a volume of 0,6 m^3?
7 Ethylene glycol antifreeze has a specific gravity of 1,06. This implies that any volume of ethylene glycol weighs 1.06 times as much as an equal volume of water. The density of water is 1×10^3 kg/m^3. What is the weight of a liter of ethylene glycol?
8 A wall 8,5 m long and 2,6 m high is to be plastered to an average thickness of 10 mm. The density of the plaster to be used is 2400 kg/m^3. What is the weight of plaster involved?

9 A construction hoist handles a maximum load of 300 kg. A wheelbarrow weighs 35 kg, and bags of cement weigh 45 kg. What is the maximum number of bags which may be loaded onto a wheelbarrow and hoisted at one time?

10 A copper pipe 3 m long has an outside diameter of 10 mm and a wall thickness of 2 mm. The density of copper is 8,9 Mg/m^3. What is the weight of the pipe?

11 (See Problem 17-13) In the forest industry one cunit of solid bone-dry wood is estimated at 2500 lb, with other materials having different multipliers: shavings, $\times$ 0.5; chips, $\times$ 0.7; sawdust, bark, and other "fines", $\times$ 0.8. Using your answer to Problem 17-13, determine weights, in kilograms, of your metric equivalent cunit.

CHAPTER NINETEEN

Angle Calculations in SI Units

Angle Units radians
degrees
minutes
seconds
steradians

With modern slide rule techniques, two important notes are worth considering:

1 Mixed numbers, such as 2π and $\frac{3\pi}{2}$, are perfectly good expressions, which can be converted into close *approximations* if they are required.
2 Decimal subdivisions of degrees are easier to read off the slide rule, and to interpolate, than are minutes and seconds.

PROBLEMS

1 $180° = \pi^r$. Convert the following degree measurements to radians, giving each answer *a*) with π, and
b) as an approximation.

a) 60°
b) 45°
c) 20°
d) 90°
e) 120°
f) 225°

2 Convert the following radian measurements to angles:

a) $2\pi^r$
b) $\frac{\pi^r}{6}$
c) $\frac{3\pi^r}{2}$
d) $\frac{5\pi^r}{6}$
e) 1^r
f) $0{,}257^r$

3 60′ = 1,00°. Convert the following degree-and-minute measurements to decimal form:

a) 45′
b) 20′
c) 20°12′
d) 65°24′
e) 104°30′
f) 1′

4 60″ = 1′. Convert the following measurements to decimal form:

a) 36″
b) 5″
c) 0°30′18″
d) 5°6′35″
e) 81°19′20″
f) 225°55′40″

5 An area of 4 mm^2 is marked out on the surface of a sphere 6 mm in diameter. What is the solid angle at the center of the sphere which subtends the 4 mm^2 area?

6 An area of 12,5 mm^2 is marked out on the surface of the same 6-mm sphere. What is the solid angle at the center?

CHAPTER TWENTY

Time Calculations in SI Units

Time Units hours
minutes
seconds
(days, weeks, etc.)

In many calculations, it may be convenient to describe time in decimal parts of hours or minutes, rather than in hours, minutes, and seconds.

PROBLEMS

1 60 min = 1,00 h. What are the decimal equivalents of the following time measurements:
a) 20 min
b) 5 min
c) 2 h 18 min
d) 5 h 40 min
e) 12 h 15 min
f) 1 min

2 60 s = 1,00 min. What are the decimal equivalents of the following time measurements:
a) 30 s
b) 55 s
c) 4 min 12 s
d) 15 min 28 s
e) 2 h 18 min 5 s
f) 15 h 35 min 40 s

3 A tradesman works for 6,75 h at $4,085 per hour. What is his income for the day?

4 A carpenter works a regular $6\frac{1}{2}$-h day at $4,32 per hour, and 2,2 h overtime at time-and-a-half. What is his income for the day?

5 A welder working alone can weld up an assembly in 6 h. His apprentice can do the same job alone in 9 h. How long should it take the two men working together?

6 How much time elapses between 9:32:08 a.m. and 2:15:26 p.m.? Answer in two forms.

Telling time in SI units utilizes the 24-hour clock which has been standard in the military service for many years. (Fig. 4-3) This system replaces 1:00 o'clock in the afternoon with 1 h past 12, or 13.00 hours. 4:00 p.m. is now 16.00 hours. 7:20 p.m. is now 19.20 hours.

7 What is the 24-hour clock designation of 3:20 a.m.?
8 What is the 24-hour clock designation of 20 minutes before 5 o'clock in the afternoon?
9 What is the "old" designation for 21.20 h?
10 How much time elapses between 08.00 h and 16.30 h?

CHAPTER TWENTY-ONE

Temperature Calculations in SI Units

Temperature Units kelvin
degree Celsius

The freezing point of pure water is 273,15 K or 0°C. (Note that kelvin temperatures do not use "degree".) A change of 1 Celsius degree is identical to a change of 1 kelvin. Most North Americans are more familiar with the Fahrenheit scale of temperature measurement, in which the freezing point of water is 32°. To convert a Fahrenheit scale reading to its equivalent Celsius reading, use the relationship

$$^{\circ}C = \tfrac{5}{9}\,(^{\circ}F - 32)$$

where C = temperature reading in °C
F = temperature reading in °F

PROBLEMS

1 Room temperatures in the range of 70°F are quite common. What is the rounded-off equivalent Celsius temperature reading?
2 "Normal" body temperature is 98,6°F. What is the equivalent Celsius thermometer reading?
3 At what temperature is the Celsius thermometer reading identical to the Fahrenheit reading?
4 What is the "size" of a Celsius degree change in terms of a Fahrenheit degree change?
5 What is the Celsius equivalent of zero kelvin?
6 What is the Celsius equivalent of 120 K?
7 What is the kelvin equivalent of 25°C?
8 What is the kelvin equivalent of 50°F?
9 What is the Fahrenheit equivalent of 20°C?
10 What is the Fahrenheit equivalent of 415 K?

CHAPTER TWENTY-TWO

Force Calculations in SI Units

Force Units newton
kilogram meter per second squared

Newton's Second Law tells us that $f = ma$

where f = force necessary to impart an acceleration to a mass, N
m = mass of material being accelerated, kg
a = acceleration imparted to mass by the force, m/s^2

One newton (N) is the force which will impart to a mass of 1 kilogram an acceleration of 1 meter per second squared.

PROBLEMS

1 A net force causes a mass of 20 kg to be accelerated at 12 m/s^2. What is the value of the force?
2 A force of 110 N is applied to a mass of 40 kg. What will be the acceleration of the 40 kg body?
3 A satellite hurtles through space with an acceleration of 65 m/s^2 under the influence of a force of 8,75 kN. What is the mass of the satellite?
4 The earth attracts a mass of 1 kg towards itself, imparting to it an acceleration of 9,81 m/s^2. What is the force of attraction impressed on the 1 kg mass?
5 A force of 1200 N acts for 12 s to change the velocity of a rocket from 636 m/s to 1860 m/s. What is the mass of the rocket?

CHAPTER TWENTY-THREE

Construction Calculations in SI Units

Length Units	meter
	millimeter
Area Units	square meters
Volume Units	cubic meters

Chapters 15 through 18 have given you some practice in working with the common SI units for construction. You should now be ready to attack the following problems.

PROBLEMS

1–5 Use your personal metric ruler to determine the lengths, in millimeters, of the dimensions shown in Figure 23-1.

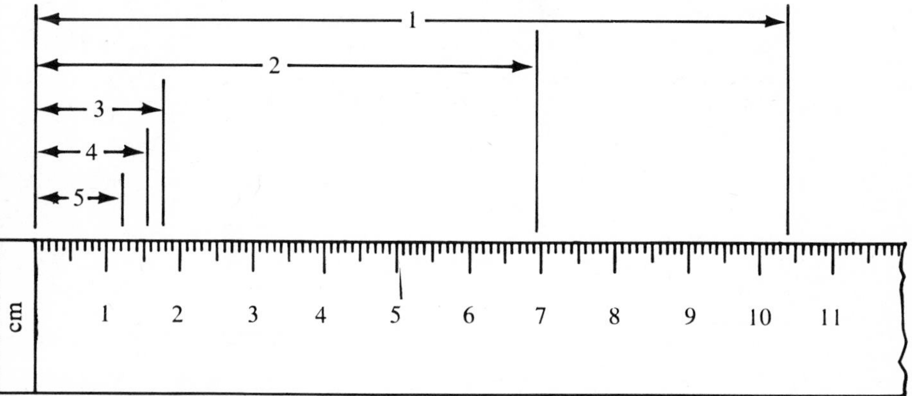

Fig. 23-1 Measure the 5 indicated dimensions in millimeters.

6 What is the outside diameter of the washer in Figure 23-2? What is the diameter of the hole?

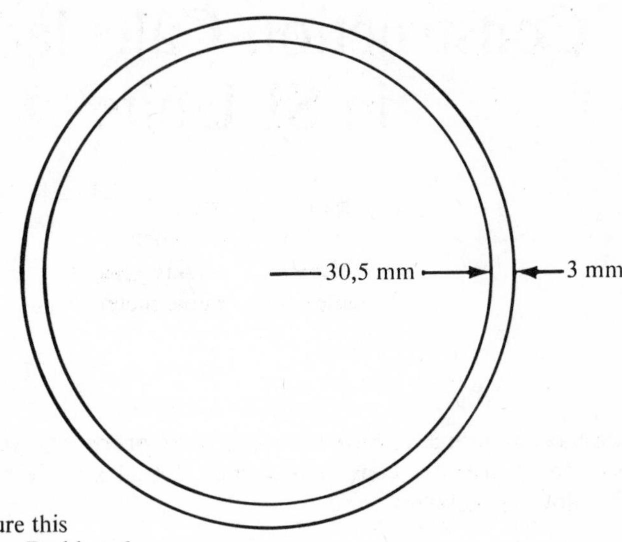

Fig. 23-2 Measure this washer: Problem 6.

7 How far will the nail in Figure 23-3 (hammered flush) protrude through a plank 47 mm thick?

Fig. 23-3 Nail for Problem 7.

8 What are the dimensions of the block of wood in Figure 23-4 in millimeters?

9 Describe the three measurements of Problem 8 in centimeters and meters.

Fig. 23-4 Wooden block for Problem 8.

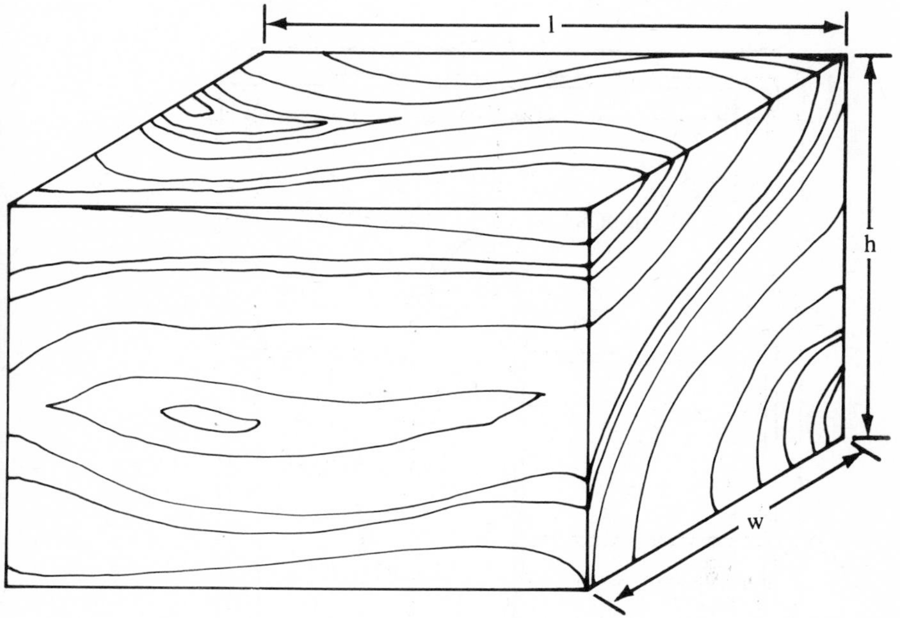

10 A customer wants a piece of lumber 2,56 m long, 3 cm wide, and 0,8 cm thick. What are these dimensions in millimeters?

11 Another customer wants a sidewalk to be 1 m 220 mm wide. Write this width in meters (which is a better way!)

12 The sidewalk in Problem 11 is to be 7 cm thick. What is this thickness in meters? in millimeters?

13 What is the volume per meter length of sidewalk in Problems 11 and 12?

14 From the floor layout in Figure 23-5,

a) Bedroom 2 is shown as 3,6 wide. What is the actual width of this room in meters? in millimeters?

b) What is the area of the living room floor space?

c) What is the total area of the floor plan?

d) If the ceilings are 2,5 m above the floor, what is the volume of the bathroom?

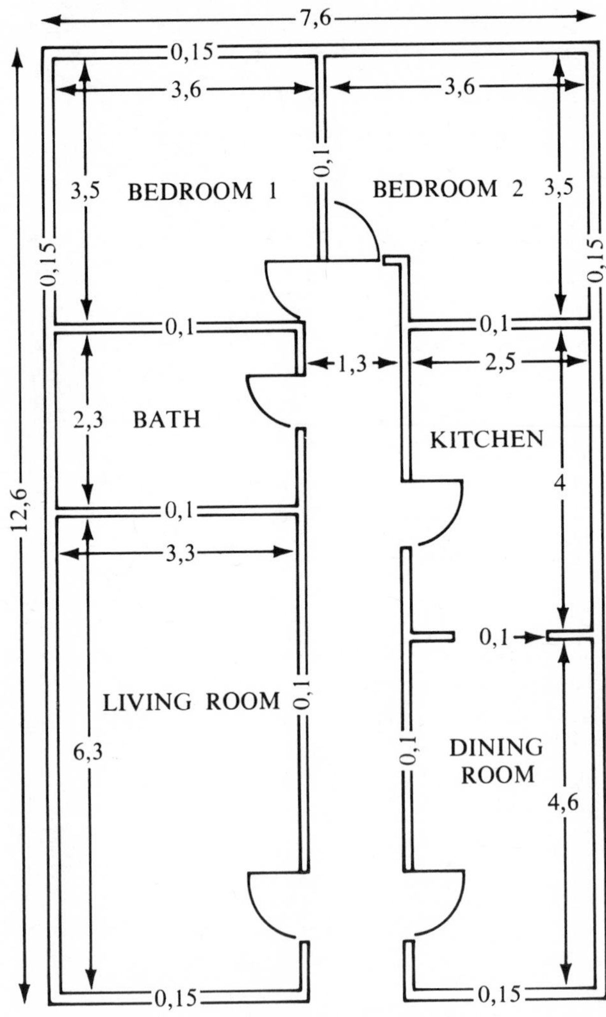

Fig. 23-5 House floor plan for Problem 14.

15 You are to paint the dining room shown in Figure 23-6. Omitting windows and portions of wall above doorways, what is the surface area of

a) ceiling?
b) walls?

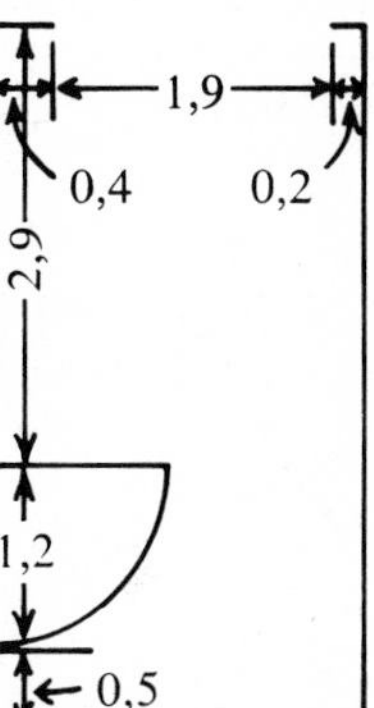

Fig. 23-6 Dining room for Problem 15.

CHAPTER TWENTY-FOUR
Heating Calculations in SI Units

Temperature Units	kelvin
	degree Celsius

Temperature in °C = K − 273,15
Temperature in °F = 1.8°C + 32
Temperature difference in kelvins = temperature difference in degrees celsius

Heat units	Quantity of heat	joule
	Heat flow rate	watt (joule per second)
	Heat capacity	joules per kelvin
	Combustion intensity	watts per cubic meter
	Heat flux	watts per square meter
	Heat transfer coefficient	watts per square meter per kelvin
	Specific energy	joules per kelvin
	Specific heat capacity	joules per kilogram per kelvin

Measurement units for various fuels:

Coal	tonne
Electricity	kilowatthour
Gas	100 megajoules
Oil	liter, tonne, cubic meter, kilogram

Older systems of heat measurement have involved British Thermal Units (BTUs) and calories and kilocalories. One of the problems has been the variety of interpretation put on calories by different professional groups. In addition to the various units shown above, there are many more used by the heating, air conditioning and refrigeration industry, including a variety of relationships involving viscosity of oils and other physical properties.

The International conversion units for heating measurements are:

1 BTU = 1 055,06 J
1 kWh = 3,6 MJ
1 therm = 105,506 MJ

PROBLEMS

1 A scientist heats an experiment from a low temperature of 220 K to a high of 310 K.
   a) What is the temperature change in kelvins?
   b) What is the temperature change in celsius degrees?
   c) What are the two temperatures in degrees Celsius?

2 A steam boiler is fired with oil having a calorific value of 35 MJ/kg. What is the equivalent energy content of a kilogram of the oil, expressed in kilowatthours? (1 watt = 1 joule per second)

3 Two furnace oil quotations are received. While the price per kiloliter is the same for each, oil *A* is specified as having a calorific value of 19 230 BTU/lb, while fuel *B* is rated at 44,82 MJ/kg. Which oil is the better buy?

4 The quantity of heat flow from a region of high temperature to a region of lower temperature through a material is given by

$$Q = \frac{kA(\Delta T)}{t} \text{ watts}$$

where k = the heat conductivity of the material, W/m^2•deg C
A = cross-sectional area of material, m^2
ΔT = temperature change, deg C
t = thickness of material, m

A 250 mm hollow tile wall with stucco exterior has a published transmission coefficient of 1,362 782 4 joules per square meter second degree celsius. Since a watt is one joule per second, what is the heat loss in watts through a cross-section of one square meter, if the inside temperature is 22°C and the outside temperature is 5°C?

CHAPTER TWENTY-FIVE

Electrical Calculations in SI Units

Electrical Units	ampere
	coulomb
	volt
	ohm
	watt
	kilowatt-hour
	farad
	henry
	hertz

PROBLEMS

1 Ohm's Law tells us that the current flow in amperes is directly proportional to the applied electromotive force in volts and inversely proportional to the resistance in ohms.

$$I = \frac{E}{R}$$

What is the current flow when a 50 volt emf is applied to a 20 kilohm circuit resistance?

2 What circuit resistance will limit the current flow to 150 μA when the applied emf is 30 V?

3 What is the voltage drop across a 3,9 kΩ resistance when the current flow through it is 2 mA?

4 Power is the product of volts and amperes. What is the power delivered to the 20 kΩ resistance in Problem 1?

5 What power is drawn from the energy source when 2,1 A flows under an emf of 110 V?

6 What is the total energy absorption if the circuit of Problem 5 is in operation for 18 h?

7 What is the frequency of a signal in hertz if the period of 1 cycle is 5 μs?

8 The capacitance of a parallel-plate capacitor is found from

$$C = \frac{0.0885\epsilon_r A(N-1)}{t} \text{ pF}$$

where ϵ_r = dielectric constant relative to air
A = area of one side of one plate in square centimeters
N = number of plates
t = dielectric thickness in centimeters

Determine an equivalent formula for capacitance in picofarads when plate area is given in square meters and dielectric thickness is given in millimeters.

9 What is the capacitance of a capacitor consisting of 2 plates each 0,5 m square, separated by 2 mm of glass of dielectric constant 3,78?

10 The capacitive reactance of a capacitor is found from

$$X_C = \frac{1}{2\pi fC}\,\Omega$$

where f = frequency at which capacitor is operating, Hz
C = capacitance, F

What is the reactance of a 47 pF capacitor at a frequency of 1,32 MHz?

11 The inductance of a single-layer air-cored solenoid is found from

$$L = \frac{n^2r^2}{9r + 10l}\,\mu\text{H}$$

where n = number of turns
r = coil radius, inches
l = coil length, inches

Develop a formula for L in microhenries when r and l are both given in millimeters.

12 What is the inductance of a 12 mm long coil consisting of 12 turns, if the coil diameter is 4 mm?

13 The inductive reactance of an inductor is found from

$$X_L = 2\pi fL\,\Omega$$

where f = frequency at which inductor is operating, Hz
L = inductance, H

What is the reactance of a 3 μH coil at a frequency of 980 kHz?

14 The resonant frequency of a circuit containing inductance and capacitance is found from

$$f_r = \frac{1}{2\pi\sqrt{LC}}\ \text{Hz}$$

where L = inductance, H
C = capacitance, F

What is the resonant frequency of a circuit containing a 22 pF capacitor and an 8 μH inductor?

CHAPTER TWENTY-SIX

Lighting Calculations in SI Units

Lighting Units candela
lumen
lux

PROBLEMS

1 Luminous flux is measured in lumens, where

$$1 \text{ lumen} = 1 \text{ cd/sr}$$

What is the luminous flux when a luminous intensity of 20 candela is generated within a solid angle of 0,4 sr?

2 What must be the solid angle enclosing a light source capable of generating 35 candela if it is to be rated at 140 lm?

3 Illumination on surfaces is measured in lux, where

$$1 \text{ lux} = 1 \text{ lm/m}^2 = 1 \text{ cd/sr}\cdot\text{m}^2$$

A lighting fixture delivers 20 klm to a surface area of 15 m^2. What is the illumination rating of the fixture in lux?

4 An intensity of 850 lux falls upon an area of 3 m^2. What is the radiated luminous flux?

5 The brightness of the surface of a light source is indicated in candelas per square meter. A recommended brightness for a fixture is about 12 lm/m^2. If the area is 1,5 m^2 and the source angle is 1 sr, what must be the luminous intensity in candelas?

CHAPTER TWENTY-SEVEN

What to Expect Along with SI Units

The great era of international cooperation which has produced SI Units will also usher in other cooperative simplifications which we should watch for and encourage. Some of these have been mentioned in preceding chapters, and a few will be introduced now. You may find it entertaining to watch for the appearance of these additional features. In some cases you may wish to discuss them with your suppliers, purchasing agents, printers, retailers, or employee and management groups.

27-1 PAPER

The international standard size of paper is a sheet whose area is 1 m^2, which is capable of being cut in half repeatedly, to give smaller sheets whose dimensions are all in the same ratio. This is achieved by making the A0 sheet 841 × 1189 mm. Notice that $1189 = 841\sqrt{2}$. When the A0 sheet is cut in half, its area will be 0.5 m^2; the former short side, 841 mm, becomes the long side, while the new short side becomes 594 mm ($= 841/\sqrt{2}$). Table 27-1 lists the more common sizes of paper in the International series. (Two double pages of *The Globe and Mail* form an A0 sheet; one double page is A1, and a single page is A2 size.)

TABLE 27-1

INTERNATIONAL PAPER SIZES

Designation	*Dimensions in mm*		*Area*
A0	841 × 1189	Basic size	1 m^2
A1	594 × 841		0,5 m^2
A2	420 × 594	Newspaper	0,25 m^2
A3	297 × 420		0,125 m^2
A4	210 × 297	Letter	6240 mm^2
A5	148 × 210	Book	3120 mm^2
A6	105 × 148	Postcard	1560 mm^2
A7	74 × 105		780 mm^2
A8	52 × 74		380 mm^2
A9	37 × 52		190 mm^2
A10	26 × 37	Postage stamp	96 mm^2
A11	18 × 26		48 mm^2
A12	13 × 18		23 mm^2

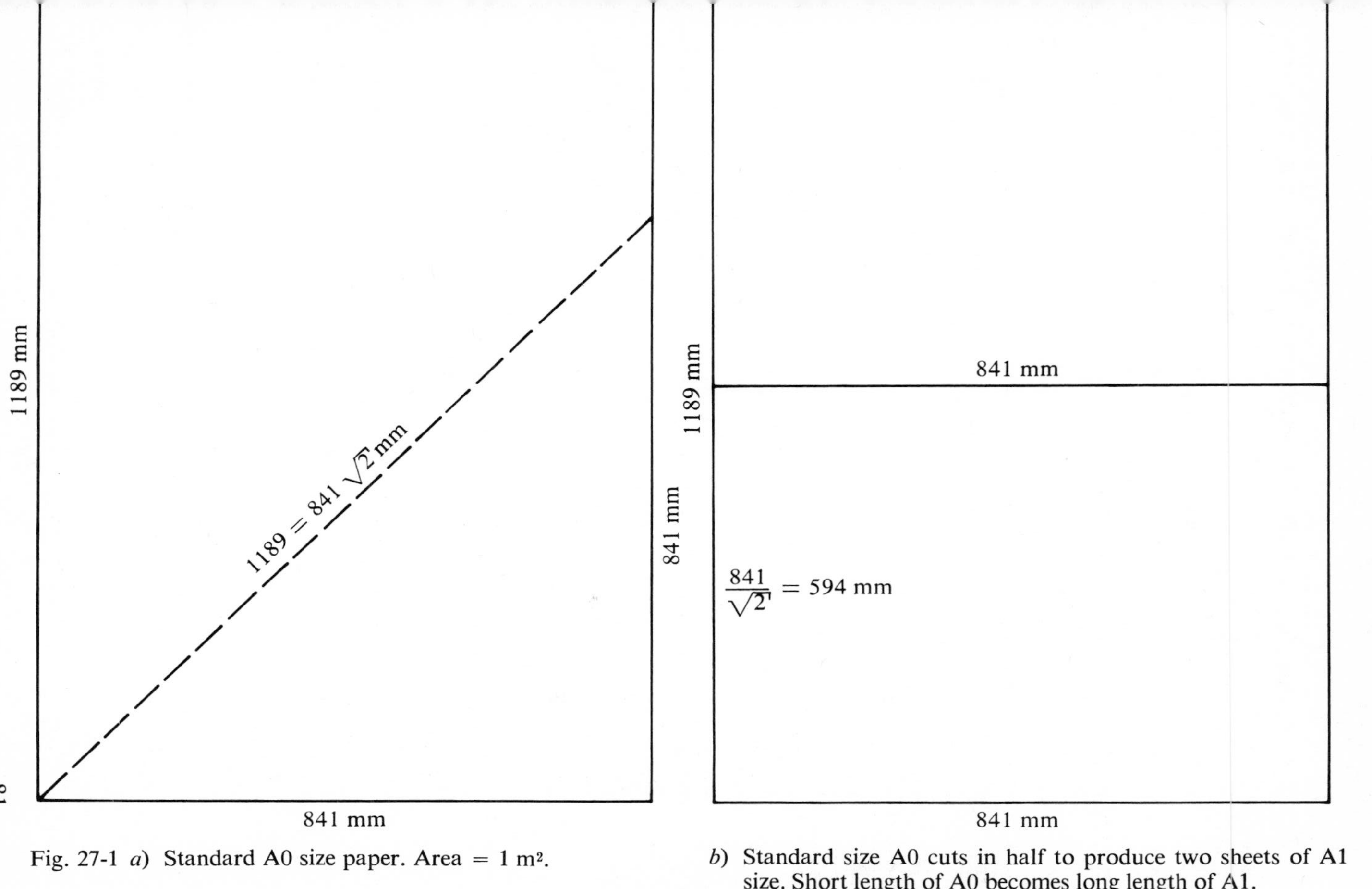

Fig. 27-1 *a*) Standard A0 size paper. Area = 1 m^2.

b) Standard size A0 cuts in half to produce two sheets of A1 size. Short length of A0 becomes long length of A1.

PROBLEMS 27-1

1 Standard filing card sizes are (a) 3 × 5 in., and (b) 5 × 7 in. What International paper sizes could replace these?

2 A popular size for photographic enlargements is 5 × 8 in. What International size could replace this?

3 Business cards generally measure about 2 × 3.5 in. What International size will probably replace this?

4 A "sheet" of toilet tissue measures $4\frac{3}{16} \times 4\frac{7}{16}$ in. What International size will probably replace this?

5 Think about the other paper sheets and forms you will probably be involved with. What sizes do you think they will become? Might some of them be reasonably established at simple fractions of the official sizes?

Paperback books	Magazines
Maps (will they fold easily?)	Catalogs
Invoices	Receipts
Prescriptions	Note books
Memo pads	Calendars
Baggage labels	Mail stickers
Admission tickets	Gift wrapping
Place mats	Facial tissues
Sandpaper	Cheques
Paper money	Membership cards

6 Measure the present spacing of the holes in the loose leaf refills of your three-ring binders. What do you think the metric spacing might be?

7 Typists and printers often describe the "weight" of their various papers as "18 Sub" or "20 Sub". What does this really signify? What will be the equivalent expressions for metric paper sizes?

8 This book has been produced in Standard paper size (although the "trim size" may vary a millimeter or two from copy to copy). What is the nominal size of the book?

27-2 SPELLING

Do you often go to the theat*er*? Or do you prefer to attend the theat*re*? Along with measurement units, efforts are being made to make spelling more universal. Some people think we should measure length in met*re*s but measure voltage with voltmet*er*s. Many hours have been spent by researchers checking dictionaries and newspapers to compare official spelling and actual practice, and point out that a presbyt*er* may have goit*re*. You may expect to see drawn-out, wordy debates in the Letters to the Editor on this subject.

Such letters have already appeared, reminding Canadians that the law defines the "kilogramme", not the "kilogram". However, the English, in the spring of 1972, revised the official spelling to the simpler form, and it is reasonable to expect that both Canada and the United States will follow their lead.

27-3 TYPEWRITER KEYBOARDS

Along with the other areas of international standardization, concerted efforts are being made to standardize typewriter keyboards. Keyboards on computer peripherals are now pretty well agreed upon, and meetings are being held from time to time on office typewriter systems. (Many attempts have been made in the past to effect standardization, but these have usually been within language groups, or in favor of special minorities. An article in the *Hamilton Spectator* on 9 Dec. 71 suggests that the 40-year old *Dvorak* proposal considers only English, and favors left-handed people.) Figure 27-2 helps to illustrate the total problem, which includes the need for a general standardization, plus the availability of some "flexible" keys to meet the need for special language requirements: Ä, Ñ, Æ, ø, etc. Figure 27-3 represents a layout of keys for English, proposed by Mr. Albert J. Mettler, Secretary of the Canadian Metric Association. Note his proposed "dead" key, drawn square, which will type the diacritical marks without moving the carriage.

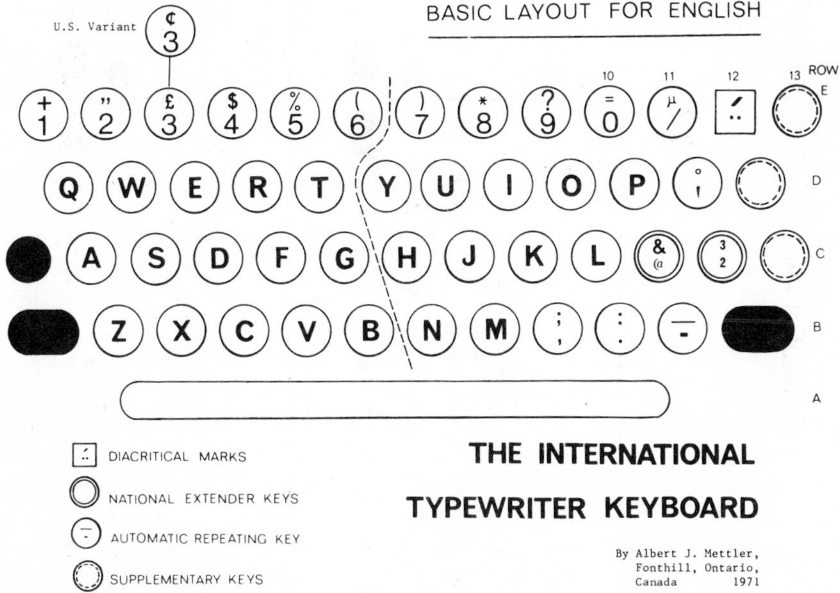

Fig. 27-2 Proposed international typewriter keyboard. *Courtesy of Mr. Albert J. Mettler, Secretary, Canadian Metric Association.*

TRADITIONAL TYPEWRITER KEYBOARDS

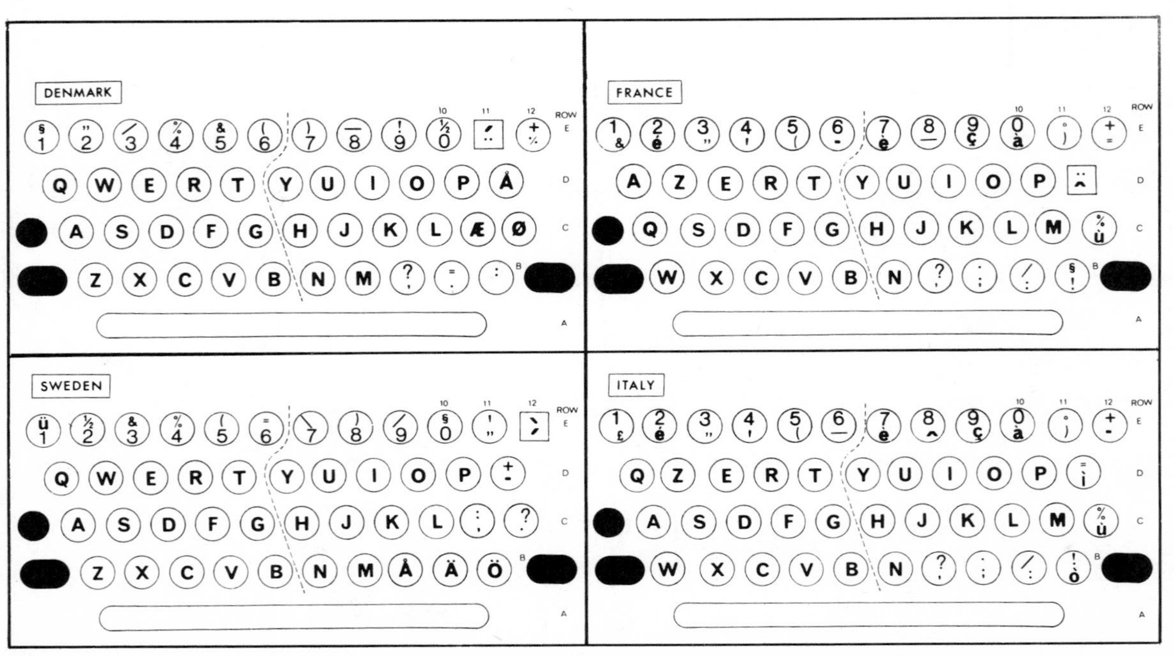

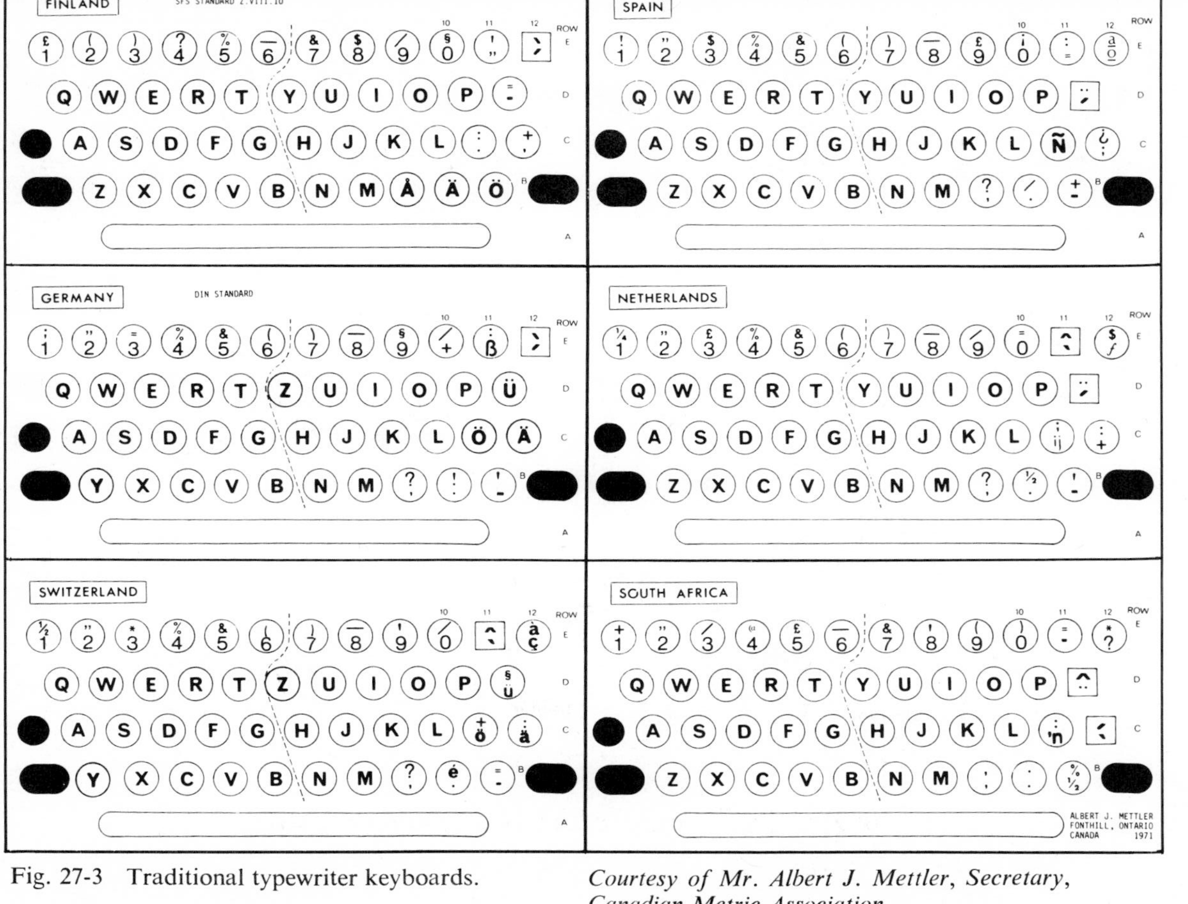

Fig. 27-3 Traditional typewriter keyboards.

Courtesy of Mr. Albert J. Mettler, Secretary, Canadian Metric Association.

27-4 DATES AND TIME

Most Canadians now use a logical system of writing dates in the sequence day-month-year, and have no problem interpreting numerical sequences: 4/3/37 must mean the fourth day of March, 1937. Americans, however, use a reverse sequence: month-day-year. An alternative notation has recently been proposed by the International Organization for Standardization, and will in all likelihood be adopted by many countries. This proposal writes dates in the sequence year-month-day, and the date shown above would be written 1937-03-04. Watch for official announcements of adoption of this proposal by Canada if you are concerned. Do not expect any early revision of the calendar, however. Months of 28, 30, and 31 days will be with us for a long time to come.

In reverse sequence of the present Canadian custom, but in direct sequence with the new ISO proposal is the writing of smaller time units in the sequence of hour-minute-second, using the 24-hour system. 14.20.10 would represent twenty minutes and ten seconds after 2 p.m.

27-5 CONTAINER SIZES

The next time you look on your kitchen, storeroom, or toolroom shelves, count the different sizes and shapes of bottles, boxes, and tins. How many different shelf spacings are needed to accommodate the various containers? Perhaps some luxury items should be packaged in fancy shapes and sizes, but watch for more suitable standardization as consumer groups, manufacturers, processors, and retailers discover the advantages of uniform standard sizes and weights.

27-6 CLOTHING SIZES

Have you ever wondered how your ten-year-old son could possibly require a size sixteen jacket? Under pressure from consumer groups, size standardization in Canadian children's clothing has come a long way, but imported clothing still keeps us guessing. Watch for gradual improvement in the sizing of clothing, and don't be too surprised when the men in your family start coming home with size 300 shoes!

27-7 SCREW THREADS

Unified, National Coarse, National Fine, Whitworth, British Standard Fine, Acme, British ISO Metric Coarse and Fine, SI Metric Coarse and Fine . . . the list is almost endless when it comes to considering the screw thread systems still in operation. The number of threads per inch (per centimeter), pitch diameter, depth of thread, angular shape of tooth, dimensions of slots or hexagonal heads, the measurements of mating nuts, all contribute to the complexities of design and standardization.

In an effort to arrive at an acceptable compromise, the American National Standards Institute has proposed the OMFS – Optimum Metric Fastener System – as a basically simpler and better fastener system.

Which system is finally added to the North American standard package remains to be determined. If you are concerned with the subject, you should study the publications of ANSI, and read the articles in the journals of the mechanical trades.

27-8 WHERE TO TURN FOR GUIDANCE IN METRICATION

This little book can only hope to introduce the subject of metrication. Some of its readers will never care to learn any more about the subject, but most North Americans will find it necessary and desirable to make a complete switch in their daily measuring operations and "quantity thinking", and "go metric" completely. Where can you learn more about the metrication and standardization subjects that will affect you? Following are a few of the information sources which could best meet your needs.

1 The Canadian Metric Association. This is a private association of businesses and individuals formed to promote metrication in Canada. The Secretary is Mr. Albert J. Mettler, P.O. Box 35, Fonthill, Ont. The Association has published a number of useful papers concerning important aspects of metrication.

2 The Standards Council of Canada, Center Building, 300 Sparks St., Ottawa, Ont. K1R 7S3, Secretary, Mr. H. E. C. Price. This coordinating body was established by Parliament in October, 1970. Its purpose is to help voluntary standardization groups to cooperate in establishing standards, and in working with international standards groups. It is also responsible for encouraging preparations for the change to the metric system, in order to achieve the best possible results at minimum cost, and to update existing standards, where appropriate, when rewriting them into metric terms. The Council is composed of representatives from government and from national business, labor, and academic organizations.

3 The Metric Commission, Box 4000, Ottawa, Ont., K1S 5G8. Chairman, Mr. S. M. Gossage. This Commission is responsible for all stages of preparing the country and its organizations and individuals for metrication, and for coordinating and providing leadership in the overall program for conversion.

The Commission publishes a newsmagazine, the METRIC MONITOR.

4 Canadian Standards Association, 178 Rexdale Boulevard, Rexdale 603, Ont. This old-time body has produced many useful papers on the subject of metrication, including CSA Standard Z234.1-1970: CONVERSION OF CANADIAN TO INTERNATIONAL METRIC UNITS.

5 The Metric Association, Inc., 2004 Ash St., Waukegan, Illinois, 60085.

Members of the Canadian Metric Association receive copies of the newsletter of this American organization, which is also a private group of businesses and individuals.

6 The American National Metric Council, 1625 Massachusetts Ave., N.W., Washington, D.C. 20036. This is a non-advocate national coordinating center for voluntary metric conversion activities in the public sector. The Council provides a resource center, and publishes the METRIC REPORTER.

7 American National Standards Institute, Inc., 1430 Broadway, N.Y. 10018. This is a private cooperative group correlating the standards work of most of the professional groups in the United States. A wide variety of technical papers are available, including the various international standards, and a five-column "metric package".

8 National Bureau of Standards, U.S. Dept. of Commerce, Washington, D.C. 20234. This is the official agency charged by Congress to officially explore conversion to metric in the U.S.A.

Of particular interest will be thirteen paperback volumes issued by the Bureau in July, 1971, and available from the Superintendent of Documents, U.S. Government Printing Office, Washington, D.C., 20402. SD numbers after the titles are the official catalog numbers.

A METRIC AMERICA: A decision whose time has come. SD: C 13.10:345

U.S. Metric Study Report: INTERNATIONAL STANDARDS SD: C 13.10:345-1

U.S. Metric Study Interim Report: FEDERAL GOVERNMENT: CIVILIAN AGENCIES SD: C 13.10:345-2

U.S. Metric Study Interim Report: COMMERCIAL WEIGHTS AND MEASURES SD: C 13.10:345-3

U.S. Metric Study Interim Report: THE MANUFACTURING INDUSTRY SD: C 13.10:345-4

U.S. Metric Study Interim Report: NONMANUFACTURING BUSINESSES SD: C 13.10:345-5

U.S. Metric Study Interim Report: EDUCATION SD: C 13.10:345-6

U.S. Metric Study Interim Report: THE CONSUMER SD: C 13.10:345-7

U.S. Metric Study Interim Report: INTERNATIONAL TRADE SD: C 13.10:345-8

U.S. Metric Study Interim Report: DEPARTMENT OF DEFENSE SD: C 13.10:345-9

U.S. Metric Study Interim Report: A HISTORY OF THE METRIC

SYSTEM CONTROVERSY IN THE UNITED STATES SD: C 13.10:345-10

U.S. Metric Study Interim Report: ENGINEERING STANDARDS SD: C 13.10:345-11

U.S. Metric Study Interim Report: TESTIMONY OF NATIONALLY REPRESENTATIVE GROUPS SD: C 13.10:345-12

9 European Agencies. While every European country has its own official bodies and standards groups, Great Britain has recently undergone most of the traumatic experiences associated with a modern-day conversion to metric, and to SI Units. Many governmental, industrial, and professional bodies in the U.K. have produced valuable publications relating to the subject. Of particular value to North Americans would be:

a) Construction Industry Training Board, Radnor House, London Road, Norbury, London S.W. 16. Ask for their CITB ADMINISTRATION GUIDE FOR METRICATION LEARNING TEXTS. They have done much to provide education in metric for all levels of employment in the construction field.

b) Ministry of Technology, P SI 2B, Room 201, Abell House, John Islip St., London S.W. 1. Their booklet THE BASES OF MEASUREMENT shows how the National Physical Laboratory maintains the British Standards against which all measurements may be checked.

c) Council of Technical Examining Bodies, c/o The City and Guilds of London Institute, 76 Portland Place, London W1N 4AA. Their pamphlet SI IN ENGINEERING gives some useful typical examination questions relating to SI Units.

d) British Standards Institution, 2 Park St., London W.1. (Sales office: 101 Pentonville Rd., London N.1). This official body has produced over 1800 metric standards for the guidance of British industry.

10 AMERICAN METRIC JOURNAL – a bimonthly magazine published by AMJ Publishing Co., Drawer L, Tarzana, California, 91356, devoted to news and articles relating to "metric training and educational progress".

11 National, State, and Provincial government departments. Most major governments now have metric committees, and some agencies are producing useful literature. One of many such pamphlets is METRICATION: A GUIDE FOR CONSUMERS published by the Canadian Government Department of Consumer and Corporate Affairs. Local public librarians and newspaper editors should be able to provide names and addresses of possible sources of documents.

THE MODERNIZED

metric system

The International System of Units-SI

is a modernized version of the metric system established by international agreement. It provides a logical and interconnected framework for all measurements in science, industry, and commerce. Officially abbreviated SI, the system is built upon a foundation of seven base units, plus two supplementary units, which appear on this chart along with their definitions. All other SI units are derived from these units. Multiples and submultiples are expressed in a decimal system. Use of metric weights and measures was legalized in the United States in 1866, and since 1893 the yard and pound have been defined in terms of the meter and the kilogram. The base units for time, electric current, amount of substance, and luminous intensity are the same in both the customary and metric systems.

COMMON CONVERSIONS
Accurate to Six Significant Figures

Symbol	When You Know	Multiply by	To Find	Symbol
in	inches	[A]25.4	[B]millimeters	mm
ft	feet	[A]0.3048	meters	m
yd	yards	[A]0.9144	meters	m
mi	miles	1.609 34	kilometers	km
yd^2	square yards	0.836 127	square meters	m^2
	acres	0.404 686	[C]hectares	ha
yd^3	cubic yards	0.764 555	cubic meters	m^3
qt	quarts (lq)	0.946 353	[D]liters	l
oz	ounces (avdp)	28.349 5	grams	g
lb	pounds (avdp)	0.453 592	kilograms	kg
°F	Fahrenheit temperature	[A]5/9 (after subtracting 32)	Celsius temperature	°C
mm	millimeters	0.039 370 1	inches	in
m	meters	3.280 84	feet	ft
m	meters	1.093 61	yards	yd
km	kilometers	0.621 371	miles	mi
m^2	square meters	1.195 99	square yards	yd^2
ha	[C]hectares	2.471 05	acres	
m^3	cubic meters	1.307 95	cubic yards	yd^3
l	[D]liters	1.056 69	quarts (lq)	qt
g	grams	0.035 274 0	ounces (avdp)	oz
kg	kilograms	2.204 62	pounds (avdp)	lb
°C	Celsius temperature	[A]9/5 (then add 32)	Fahrenheit temperature	°F

[A]exact

[B]for example, 1 in = 25.4 mm, so 3 inches would be

$(3\text{ in})(25.4\ \frac{\text{mm}}{\text{in}}) = 76.2\text{ mm}$

hectare is a common name for 10 000 square meters

[D]liter is a common name for fluid volume of 0.001 cubic meter

Note: Most symbols are written with lower case letters; exceptions are units named after persons for which the symbols are capitalized. Periods are not used with any symbols.

MULTIPLES AND PREFIXES
These Prefixes May Be Applied To All SI Units

Multiples and Submultiples		Prefixes	Symbols
1 000 000 000 000 =	10^{12}	tera (tĕr′ȧ)	T
1 000 000 000 =	10^{9}	giga (jĭ′gȧ)	G
1 000 000 =	10^{6}	mega (mĕg′ȧ)	M
1 000 =	10^{3}	kilo (kĭl′ō)	k
100 =	10^{2}	hecto (hĕk′tō)	h
10 =	10^{1}	deka (dĕk′ȧ)	da
Base Unit 1	10^{0}		
0.1 =	10^{-1}	deci (dĕs′ĭ)	d
0.01 =	10^{-2}	centi (sĕn′tĭ)	c
0.001 =	10^{-3}	milli (mĭl′ĭ)	m
0.000 001 =	10^{-6}	micro (mī′krō)	µ
0.000 000 001 =	10^{-9}	nano (năn′ō)	n
0.000 000 000 001 =	10^{-12}	pico (pē′kō)	p
0.000 000 000 000 001 =	10^{-15}	femto (fĕm′tō)	f
0.000 000 000 000 000 001 =	10^{-18}	atto (ăt′tō)	a

National Bureau of Standards
Special Publication 304A (Revised October 1972)

For sale by the Superintendent of Documents, U.S. Government Printing Office, Washington, D.C. 20402
SD Catalog No. C13.10: 304A – Price 25 cents

REFERENCES

NBS Special Publication 330, 1972 Edition, International System of Units (SI), available by purchase from the Superintendent of Documents, Government Printing Office, Washington, D.C. 20402, order as C13.10:330/2; 30 cents a copy.

ASTM Metric Practice Guide E380-72, available by purchase from the American Society for Testing and Materials, 1916 Race Street, Philadelphia, Pa. 19103, $1.50 a copy, minimum order $3.00.

Rules for the Use of Units of the International System of Units, order as ISO Recommendation R1000; $1.25 a copy from the American National Standards Institute, 1430 Broadway, New York, N.Y. 10018.

meter-m
LENGTH

kilogram-kg
MASS

second-s
TIME

ampere-A
ELECTRIC CURREN

kelvin-K
TEMPERATURE

mole-mol
AMOUNT OF SUBS

candela-cd
LUMINOUS INTENS

radian-rad
PLANE ANGLE

INCHES
1 2 3 4 5 6 7 8 9 10 11 12 13 14 15 16
10 20 30 40
CENTIMETERS

U.S. DEPARTMENT OF COMMERCE
NATIONAL BUREAU OF STANDARDS

SEVEN BASE UNITS

lling, metre)
s in vacuum of
krypton-86.

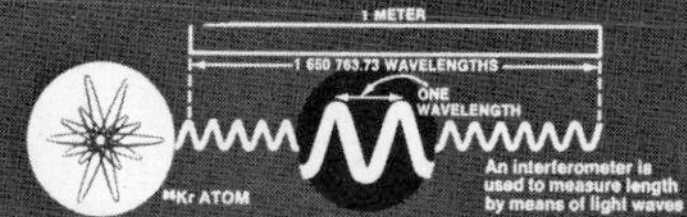

The SI unit of area is the **square meter** (m^2).

The SI unit of volume is the **cubic meter** (m^3). The liter (0.001 cubic meter), although not an SI unit, is commonly used to measure fluid volume.

it of mass, the kilogram, is a
um alloy kept by the Interna-
and Measures at Paris. A du-
the National Bureau of Stand-
tandard for the United States.
still defined by an artifact.

The SI unit of force is the **newton** (N). One newton is the force which, when applied to a 1 kilogram mass, will give the kilogram mass an acceleration of 1 (meter per second) per second.

$1N = 1kg{\cdot}m/s^2$

The SI unit for pressure is the **pascal** (Pa).
$1Pa = 1N/m^2$

The SI unit for work and energy of any kind is the **joule** (J).
$1J = 1N{\cdot}m$

The SI unit for power of any kind is the **watt** (W).
$1W = 1J/s$

of 9 192 631 770
ith a specified
is realized by
e frequency of
gh a system of
detector.

TRANSITION REGION (CAVITY) OSCILLATING FIELD
CESIUM SOURCE
DETECTOR
DEFLECTION MAGNET
DEFLECTION MAGNET
OSCILLATOR
NBS ATOMIC TIME SCALE SYSTEM

r or "clock." Only those atoms whose
n region reach the detector. When
ck indicates one second has passed.

The number of periods or cycles per second is called frequency. The SI unit for frequency is the **hertz** (Hz). One hertz equals one cycle per second.

The SI unit for speed is the **meter per second** (m/s).

The SI unit for acceleration is the **(meter per second) per second** (m/s^2).

Standard frequencies and correct time are broadcast from WWV, WWVB, and WWVH, and stations of the U.S. Navy. Many short-wave receivers pick up WWV and WWVH, on frequencies of 2.5, 5, 10, 15, and 20 megahertz.

d as that current which, if maintained in each of two
eparated by one meter in free space, would produce
two wires (due to their magnetic fields) of 2×10^{-7}
er of length.

The SI unit of voltage is the **volt** (V).
$1V = 1W/A$

The SI unit of electric resistance is the **ohm** (Ω).
$1\Omega = 1V/A$

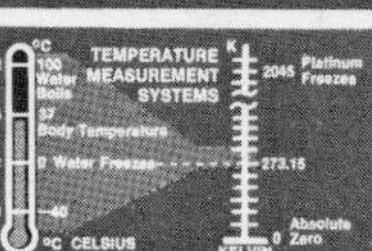

On the commonly used Celsius temperature scale, water freezes at about 0 °C and boils at about 100 °C. The °C is defined as an interval of 1 K, and the Celsius temperature 0 °C is defined as 273.15 K.

1.8 Fahrenheit degrees are equal to 1.0 °C or 1.0 K; the Fahrenheit scale uses 32 °F as a temperature corresponding to 0 °C.

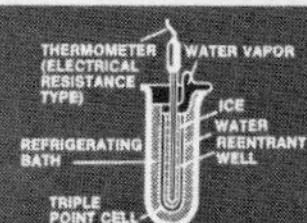

The standard temperature at the triple point of water is provided by a special cell, an evacuated glass cylinder containing pure water. When the cell is cooled until a mantle of ice forms around the reentrant well, the temperature at the interface of solid, liquid, and vapor is 273.16 K. Thermometers to be calibrated are placed in the reentrant well.

is the amount of substance of
hat contains as many elemen-
es as there are atoms in 0.012
of carbon 12.

When the mole is used, the elementary entities must be specified and may be atoms, molecules, ions, electrons, other particles, or specified groups of such particles.

The SI unit of concentration (of amount of substance) is the **mole per cubic meter** (mol/m^3).

is defined as the luminous
/600 000 of a square meter
dy at the temperature of
inum (2045 K).

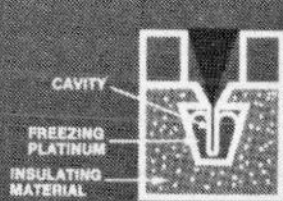

The SI unit of light flux is the **lumen** (lm). A source having an intensity of 1 candela in all directions radiates a light flux of 4 π lumens.

TWO SUPPLEMENTARY UNITS

steradian-sr SOLID ANGLE

The steradian is the solid angle with its vertex at the center of a sphere that is subtended by an area of the spherical surface equal to that of a square with sides equal in length to the radius.

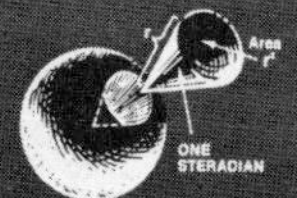

21 22 23 24 25 26 27 28 29 30 31 32 33 34 35

60 70 80 90

Conversion Tables

Example

In the conversion tables which follow, the method of conversion is as follows:

The number of		multiplied by	equal	the number of
feet	$\times$	12	=	inches

To convert 6 feet to the equivalent length in inches, multiply the number of feet (6) by 12 to obtain the number of inches ($6 \times 12 = 72$).

CONVERSION TABLE 1
LENGTH CONVERSIONS

Angstrom units	$\times 1 \times 10^{-10}$	= meters
	$\times 1 \times 10^{-4}$	= microns
	$\times 1{,}650\ 763\ 73 \times 10^{-4}$	= wavelengths of orange-red line of krypton 86
Cables	$\times 120$	= fathoms
	$\times 720$	= feet
	$\times 219{,}456$	= meters
Fathoms	$\times 6$	= feet
	$\times 1{,}828\ 8$	= meters
Feet	$\times 12$	= inches
	$\times 0{,}3048$	= meters
Furlongs	$\times 660$	= feet
	$\times 201{,}168$	= meters
	$\times 220$	= yards
Inches	$\times 2{,}54 \times 10^{8}$	= Angstroms
	$\times 25{,}4$	= millimeters
	$\times 8{,}333\ 33 \times 10^{-2}$	= feet
Kilometers	$\times 3{,}280\ 839 \times 10^{3}$	= feet
	$\times 0{,}539\ 956$	= nautical miles
	$\times 0{,}621\ 371$	= statute miles
	$\times 1{,}093\ 613 \times 10^{3}$	= yards
Light-years	$\times 9{,}460\ 55 \times 10^{12}$	= kilometers
	$\times 5{,}878\ 51 \times 10^{12}$	= statute miles
Meters	$\times 1 \times 10^{10}$	= Angstroms
	$\times 3{,}280\ 839\ 9$	= feet
	$\times 39{,}370\ 079$	= inches
	$\times 1{,}093\ 61$	= yards
Microns	$\times 10^{4}$	= Angstroms
	$\times 10^{-4}$	= centimeters
	$\times 10^{-6}$	= meters
Nautical Miles (International)	$\times 8{,}439\ 049$	= cables
	$\times 6{,}076\ 115\ 49 \times 10^{3}$	= feet
	$\times 1{,}852 \times 10^{3}$	= meters
	$\times 1{,}150\ 77$	= statute miles

CONVERSION TABLE 1 contd.

Statute Miles	$\times$ 5,280 $\times 10^3$	= feet
	$\times$ 8	= furlongs
	$\times$ 6,336 0 $\times 10^4$	= inches
	$\times$ 1,609 34	= kilometers
	$\times$ 8,689 7 $\times 10^{-1}$	= nautical miles
Mils	$\times 10^{-3}$	= inches
	$\times$ 2,54 $\times 10^{-2}$	= millimeters
	$\times$ 25,4	= micrometers
Yards	$\times$ 3	= feet
	$\times$ 9,144 $\times 10^{-1}$	= meters
Feet/hour	$\times$ 3,048 $\times 10^{-4}$	= kilometers/hour
	$\times$ 1,645 788 $\times 10^{-4}$	= knots
Feet/minute	$\times$ 0,3048	= meters/minute
	$\times$ 5,08 $\times 10^{-3}$	= meters/second
Feet/second	$\times$ 1,097 28	= kilometers/hour
	$\times$ 18,288	= meters/minute
Kilometers/hour	$\times$ 3,280 839 $\times 10^3$	= feet/hour
	$\times$ 54,680 66	= feet/minute
	$\times$ 0,277 777	= meters/second
	$\times$ 0,621 371	= miles/hour
Kilometers/minute	$\times$ 3,280 839 $\times 10^3$	= feet/minute
	$\times$ 37,282 27	= miles/hour
Knots	$\times$ 6,076 115 $\times 10^3$	= feet/hour
	$\times$ 101,268 5	= feet/minute
	$\times$ 1,687 809	= feet/second
	$\times$ 1,852	= kilometers/hour
	$\times$ 30,866	= meters/minute
	$\times$ 0,514 4	= meters/second
	$\times$ 1,150 77	= statute miles/hour
Meters/hour	$\times$ 3,280 839	= feet/hour
	$\times$ 88	= feet/minute
	$\times$ 1,466	= feet/second
	$\times$ 1 $\times 10^{-3}$	= kilometers/hour
	$\times$ 1,667 $\times 10^{-2}$	= meters/minute
Feet/second^2	$\times$ 1,097 28	= kilometers/hour/second
	$\times$ 0,304 8	= meters/second^2

CONVERSION TABLE 2
AREA CONVERSIONS

Acres	$\times$ 4,046 85 $\times$ 10^{-3}	= square kilometers
	$\times$ 4,046 856 $\times$ 10^{3}	= square meters
	$\times$ 4,356 0 $\times$ 10^{4}	= square feet
Ares	$\times$ 2,471 053 8 $\times$ 10^{-2}	= acres
	$\times$ 1	= square dekameters
	$\times$ 10^{2}	= square meters
Barns	$\times$ 1 $\times$ 10^{-28}	= square meters
Circular mils	$\times$ 1 $\times$ 10^{-6}	= circular inches
	$\times$ 5,067 074 8 $\times$ 10^{-4}	= square millimeters
	$\times$ 0,785 398 1	= square mils
Hectares	$\times$ 2,471 05	= acres
	$\times$ 10^{2}	= ares
	$\times$ 10^{4}	= square meters
Square feet	$\times$ 2,295 684 $\times$ 10^{-5}	= acres
	$\times$ 9,290 3 $\times$ 10^{-4}	= ares
	$\times$ 144	= square inches
	$\times$ 9,290 304 $\times$ 10^{-2}	= square meters
Square inches	$\times$ 1,273 239 5 $\times$ 10^{6}	= circular mils
	$\times$ 6,944 4 $\times$ 10^{-3}	= square feet
	$\times$ 6,451 6 $\times$ 10^{-4}	= square meters
Square kilometers	$\times$ 247,105 38	= acres
	$\times$ 1,076 391 0 $\times$ 10^{7}	= square feet
Square meters	$\times$ 10,763 9	= square feet
Square miles	$\times$ 640	= acres
	$\times$ 2,787 828 8 $\times$ 10^{7}	= square feet
	$\times$ 2,589 988 1	= square kilometers
Square mils	$\times$ 1,273 23	= circular mils
	$\times$ 10^{-6}	= square inches

CONVERSION TABLE 3
VOLUME CONVERSIONS

Acre-feet	$\times$ 1,233 481 4 $\times 10^{3}$	= cubic meters
Acre-inches	$\times$ 102,79	= cubic meters
Board-feet	$\times$ 2,359 737 $\times 10^{-3}$	= cubic meters
	$\times$ 144	= cubic inches
Bushels (Imperial)	$\times$ 3,636 87 $\times 10^{-2}$	= cubic meters
	$\times$ 1,284 348	= cubic feet
Bushels (US)	$\times$ 3,523 907 $\times 10^{-2}$	= cubic meters
	$\times$ 1,244 456	= cubic feet
	$\times$ 8	= US dry gallons
	$\times$ 35,238 08	= liters
Cords	$\times$ 128	= cubic feet
	$\times$ 3,624 57	= cubic meters
Cubic feet	$\times$ 2,831 6 $\times 10^{-2}$	= cubic meters
	$\times$ 28,316 85	= liters
Cubic feet of water	$\times$ 62,426 2	= pounds of water
Cubic feet per hour	$\times$ 28,316 85	= liters per hour
Cubic feet per minute	$\times$ 0,471 934	= liters per second
Cubic feet per second	$\times$ 28,316 05	= liters per second
Cubic inches	$\times$ 1,638 706 4 $\times 10^{4}$	= cubic millimeters
	$\times$ 1,638 706 4 $\times 10^{-5}$	= cubic meters
	$\times$ 1,638 7 $\times 10^{-2}$	= liters
	$\times$ 16,386 71	= milliliters
Cubic meters	$\times$ 8,107 13 $\times 10^{-4}$	= acre-feet
	$\times$ 35,314 667	= cubic feet
Cubic millimeters	$\times 10^{-3}$	= cubic centimeters
	$\times 10^{-9}$	= cubic meters
	$\times$ 6,102 374 4 $\times 10^{-5}$	= cubic inches

CONVERSION TABLE 3 contd.

Cubic yards	× 27	= cubic feet
	× 0,764 555	= cubic meters
Gallons (Imperial)	× 277,4	= cubic inches
	× 4,546 090	= liters
	× 10	= pounds of water
Gallons (US dry)	× 268,802 5	= cubic inches
	× 4,404 884	= liters
Gallons (US liquid)	× 231	= cubic inches
	× 3,785 412	= liters
Kiloliters	× 35,314 67	= cubic feet
	× 6,102 374 × 10^4	= cubic inches
	× 1,000	= cubic meters
	× 1,307 950 6	= cubic yards
	× 219,969	= imperial gallons
Liters	× 10^3	= cubic centimeters
	× 1,000 × 10^6	= cubic millimeters
	× 1,000 × 10^{-3}	= cubic meters
	× 61,023 74	= cubic inches
	× 3,531 5 × 10^{-2}	= cubic feet
	× 1,307 95 × 10^{-3}	= cubic yards
	× 0,219 969	= imperial gallons
	× 0,879 877	= imperial quarts
Imperial pints	× 0,125	= imperial gallons
	× 0,568 261	= liters
	× 20	= imperial fluid ounces
	× 0,5	= imperial quarts
	× 568,260 9	= cubic centimeters
Imperial quarts	× 1,136 52 × 10^3	= cubic centimeters
	× 69,354 8	= cubic inches
	× 1,136 522 8	= liters

CONVERSION TABLE 4
MASS CONVERSIONS

Grains	$\times$ 6,479 8 $\times 10^{-2}$	= grams
	$\times$ 2,285 71 $\times 10^{-3}$	= avoirdupois ounces
Grams	$\times$ 15,432 358	= grains
	$\times$ 3,527 396 $\times 10^{-2}$	= avoirdupois ounces
	$\times$ 2,204 62 $\times 10^{-3}$	= avoirdupois pounds
Kilograms	$\times$ 564,383 4	= avoirdupois drams
	$\times$ 2,204 622 6	= avoirdupois pounds
	$\times$ 9,842 065 $\times 10^{-4}$	= long tons
	$\times 10^{-3}$	= metric tons
	$\times$ 1,102 31 $\times 10^{-3}$	= short tons
Avoirdupois ounces	$\times$ 28,349 5	= grams
	$\times$ 6,25 $\times 10^{-2}$	= avoirdupois pounds
	$\times$ 0,911 458	= troy ounces
Avoirdupois pounds	$\times$ 256	= drams
	$\times$ 4,535 923 7 $\times 10^{2}$	= grams
	$\times$ 0,453 592 4	= kilograms
	$\times$ 16	= ounces
Long tons	$\times$ 2,24 $\times 10^{3}$	= avoirdupois pounds
	$\times$ 1,106 046 9	= metric tons
	$\times$ 1,12	= short tons
Metric tons	$\times 10^{3}$	= kilograms
	$\times$ 2,204 622 $\times 10^{3}$	= avoirdupois pounds
Short tons	$\times$ 2 $\times 10^{3}$	= avoirdupois pounds
	$\times$ 907,184 74	= kilograms

CONVERSION TABLE 5
FORCE CONVERSIONS

Dynes	$\times 10^{-5}$	= newtons
Newtons	$\times 10^{5}$	= dynes
	$\times$ 0,224 808	= pounds-force
Pounds	$\times$ 4,448 22	= newtons

CONVERSION TABLE 6
ENERGY CONVERSIONS

British Thermal Units (thermochemical)	$\times$ 1,054 35 $\times 10^{3}$	= joules
	$\times$ 2,928 27 $\times 10^{-4}$	= kilowatthours
	$\times$ 1,054 35 $\times 10^{3}$	= wattseconds
Foot-pound-force	$\times$ 1,355 818 0	= joules
	$\times$ 0,138 255	= kilogramforce-meters
	$\times$ 3,766 16 $\times 10^{-7}$	= kilowatthours
	$\times$ 1,355 818 0	= newtonmeters
Joules	$\times$ 9,484 5 $\times 10^{-4}$	= British Thermal Units
	$\times$ 0,737 562	= foot-pounds-force
	$\times$ 0,101 971 6	= kilogramforce-meters
	$\times$ 2,777 7 $\times 10^{-7}$	= kilowatthours
	$\times$ 1	= wattseconds
Kilogramforce-meters	$\times$ 9,287 7 $\times 10^{-3}$	= British Thermal Units
	$\times$ 7,233 01	= foot-pounds-force
	$\times$ 9,806 65	= joules
	$\times$ 9,806 65	= newtonmeters
	$\times$ 2,724 0 $\times 10^{-3}$	= watthours
Kilowatthours	$\times$ 3,409 52 $\times 10^{3}$	= British Thermal Units
	$\times$ 2,655 22 $\times 10^{6}$	= foot-pounds-force
	$\times$ 1,341 02	= horsepowerhours
	$\times$ 3,6 $\times 10^{6}$	= joules
	$\times$ 3,670 98 $\times 10^{5}$	= kilogramforce-meters
Newtonmeters	$\times$ 0,101 971	= kilogramforce-meters
	$\times$ 0,737 562	= poundforce-feet
Watthours	$\times$ 3,414 43	= British Thermal Units
	$\times$ 2,655 22 $\times 10^{3}$	= foot-pounds-force
	$\times$ 3,6 $\times 10^{3}$	= joules
	$\times$ 3,670 98 $\times 10^{2}$	= kilogramforce-meters

CONVERSION TABLE 7
POWER CONVERSIONS

British Thermal Units/hour	$\times$ 2,928 7 $\times 10^{-4}$	= kilowatts
	$\times$ 0,292 875	= watts
BTU/minute	$\times$ 1,757 25 $\times 10^{-2}$	= kilowatts
BTU/pound	$\times$ 2,324 4	= joules/gram
BTU/second	$\times$ 1,413 91	= horsepower
	$\times$ 107,514	= kilogrammeters/second
	$\times$ 1,054 35	= kilowatts
	$\times$ 1,054 35 $\times 10^{3}$	= watts
Foot-pound-force /hour	$\times$ 5,050 $\times 10^{-7}$	= horsepower
	$\times$ 3,766 16 $\times 10^{-7}$	= kilowatts
Foot-pound-force /minute	$\times$ 3,030 303 $\times 10^{-5}$	= horsepower
	$\times$ 2,259 70 $\times 10^{-2}$	= joules/second
	$\times$ 2,259 70 $\times 10^{-5}$	= kilowatts
Horsepower	$\times$ 42,435 6	= BTU/minute
	$\times$ 550	= footpounds/second
	$\times$ 0,746	= kilowatts
	$\times$ 746	= joules/second
Kilogrammeters /second	$\times$ 9,806 65	= watts
Kilowatts	$\times$ 3,414 43 $\times 10^{3}$	= BTU/hour
	$\times$ 2,655 22 $\times 10^{6}$	= footpounds/hour
	$\times$ 4,425 37 $\times 10^{4}$	= footpounds/minute
	$\times$ 737,562	= footpounds/second
	$\times$ 1,019 726 $\times 10^{7}$	= gramcentimeters/second
	$\times$ 1,341 02	= horsepower
	$\times$ 3,6 $\times 10^{6}$	= joules/hour
	$\times 10^{3}$	= joules/second
	$\times$ 3,671 01 $\times 10^{5}$	= kilogrammeters/hour
	$\times$ 999,835	= international watt
Watts	$\times$ 44,253 7	= footpounds/minute
	$\times$ 1,341 02 $\times 10^{-3}$	= horsepower
	$\times$ 1	= joules/second

CONVERSION TABLE 8
TIME CONVERSIONS

(No attempt has been made in this brief treatment to correlate solar, mean solar, sidereal, and mean sidereal days.)

Mean solar days	$\times$ 24	= mean solar hours
Mean solar hours	$\times$ 3,600 $\times$ 10^3	= mean solar seconds
	$\times$ 60	= mean solar minutes

CONVERSION TABLE 9
ANGLE CONVERSIONS

Degrees	$\times$ 60	= minutes
	$\times$ 1,745 329 3 $\times$ 10^{-2}	= radians
Degrees /foot	$\times$ 5,726 145 $\times$ 10^{-4}	= radians /centimeter
Degrees /minute	$\times$ 2,908 8 $\times$ 10^{-4}	= radians /second
	$\times$ 4,629 629 $\times$ 10^{-5}	= revolutions /second
Degrees /second	$\times$ 1,745 329 3 $\times$ 10^{-2}	= radians /second
	$\times$ 0,166	= revolutions /minute
	$\times$ 2,77 $\times$ 10^{-3}	= revolutions /second
Minutes	$\times$ 1,667 $\times$ 10^{-2}	= degrees
	$\times$ 2,908 8 $\times$ 10^{-4}	= radians
	$\times$ 60	= seconds
Radians	$\times$ 0,159 154	= circumferences
	$\times$ 57,295 77	= degrees
	$\times$ 3,437 746 $\times$ 10^3	= minutes
Seconds	$\times$ 2,777 $\times$ 10^{-4}	= degrees
	$\times$ 1,667 $\times$ 10^{-2}	= minutes
	$\times$ 4,848 136 8 $\times$ 10^{-6}	= radians
Steradians	$\times$ 0,159 154 9	= hemispheres
	$\times$ 7,957 74 $\times$ 10^{-2}	= spheres
	$\times$ 0,636 619 7	= spherical right angles

CONVERSION TABLE 10
PRESSURE CONVERSIONS

Atmospheres	× 1,013 25	= bars
	× 1,033 23 × 10^3	= grams/square centimeter
	× 1,033 23 × 10^7	= grams/square meter
	× 14,696 0	= pounds/square inch
	× 760	= torrs
	× 101	= kilopascals
Bars	× 0,986 923	= atmospheres
	× 10^6	= baryes
	× 1,019 716 × 10^7	= grams/square meter
	× 1,019 716 × 10^4	= kilogramsforce/square meter
	× 14,503 8	= poundsforce/square inch
Baryes	× 10^{-6}	= bars
Inches of mercury	× 3,386 4 × 10^{-2}	= bars
	× 345,316	= kilogramsforce/square meter
	× 70,726 2	= poundsforce/square foot
Pascal	× 1	= newton/square meter

CONVERSION TABLE 11
ELECTRICAL CONVERSIONS

Amperes	× 1	= coulombs/second
Amperehours	× 3,6 × 10^3	= coulombs
Bels	× 10	= decibels
Coulombs	× 6,241 96 × 10^{18}	= electron charges
Megmhos/centimeter	× 2,54	= megsiemens/inch cube
Megmhos/inch	× 0,393 700 79	= megsiemens/centimeter
Nepers	× 8,686	= decibels
Mhos	× 1	= siemens

CONVERSION TABLE 12
LIGHTING CONVERSIONS

International candela	× 1	= lumens/steradian
Footcandles	× 1	= lumens/square foot
	× 10,763 9	= lumens/square meter (lux)
Footlamberts	× 1	= lumens/square foot
Lamberts	× 0,318 30	= candles/square centimeter
	× 295,719	= candles/square foot
	× 2,053 60	= candles/square inch
	× 929,03	= footlamberts
	× 1	= lumens/square centimeter
Lumens	$\times$ 7,957 74 $\times 10^{-2}$	= spherical candlepower
Lumens/square centimeter	× 1	= lamberts
Lumens/square centimeter steradian	× 3,141 59	= lamberts
Lumens/square foot	× 1	= footcandles
	× 1	= footlamberts
	× 10,763 910	= lumens/square meter
Lumens/square meter	$\times$ 9,290 30 $\times 10^{-2}$	= footcandles
Lux	$\times$ 9,290 30 $\times 10^{-2}$	= footcandles
	× 1	= lumens/square meter
Phots	$\times 10^4$	= lux

CONVERSION TABLE 13
MAGNETIC CONVERSIONS

Gausses	$\times$ 1	= lines/square centimeter
	$\times$ 1	= maxwells/square centimeter
Gilberts	$\times$ 0,795 774 72	= ampereturns
Kilolines	$\times 10^3$	= maxwells
	$\times 10^{-5}$	= webers
Lines	$\times$ 1	= maxwells
Lines/square centimeter	$\times$ 1	= gausses
Lines/square inch	$\times$ 0,155	= gausses
	$\times 10^{-8}$	= webers/square inch
Maxwells	$\times$ 1	= lines
Maxwells/square centimeter	$\times$ 6,451 6	= maxwells/square inch
Oersteds	$\times$ 2,021 267	= ampereturns/inch
	$\times$ 125,7	= ampereturns/meter
	$\times$ 1	= gilberts/centimeter
Tesla	$\times$ 1	= webers/square meter
	$\times 10^4$	= gausses
Webers	$\times 10^8$	= lines
Webers/square centimeter	$\times 10^8$	= gausses
	$\times 6{,}451\,6 \times 10^8$	= lines/square inch

TABLE 14
ABBREVIATIONS

Only the most common abbreviations have been included in this little table. For compound abbreviations other than the most ordinary, consult the individual components of the abbreviation.

Word or phrase	*Abbreviation*
Ampere	A
atto (= 10^{-18})	a
avoirdupois	avdp
barn	b
British Thermal Unit	BTU
candela	cd
centi	c
coulomb	C
cube, cubic	$\ldots^3$
curie	Ci
cycle per second	Hz
deca *or* deka	dk (da)
deci (= 10^{-1})	d
degree	$\ldots^{\circ}$
degree Celsius	°C
degree change	deg
degree Fahrenheit	°F
degree Kelvin	K
electronvolt	eV
farad	F
femto (= 10^{-15})	f
foot	ft
gallon	gal
giga (= 10^9)	G
gram	g
hectare	ha
hecto (= 10^2)	h
henry	H
hertz	Hz
hour	h

Word or phrase	*Abbreviation*
inch	in.
joule	J
kilo (= 10^3)	k
kilogram	kg
kilometer	km
kilowatthour	kWh
knot	kn.
liter	ℓ, *or* spell
lumen	lm
lux	lx
mega (= 10^6)	M
meter	m
micro	μ
microampere	μA
milli (= 10^{-3})	m
milliampere	mA
milligram	mg
millimeter	mm
minute	min
nano (= 10^{-9})	n
newton	N
ohm	Ω
parsec	pc
pascal	Pa
per	. . ./. . .
pico (= 10^{-12})	p
pint	pt
poise	P
pound	lb
quart	qt
radian	rad *or* $\ldots^{r}$
revolution	rev
revolutions per minute	rev/min

Word or phrase	*Abbreviation*
second	s
siemens	S
square	$\ldots^2$
square meters	m^2
steradian	sr
stokes	St
tera (= 10^{12})	T
tesla	T
tonne	t
voltage	V, E
watt	W
weber	Wb

Answers to Problems

ANSWERS, Chapter 14

14-1

1	124 m 270 mm
2	18 m 15 mm
3	6 kg 193 g
4	535 ml
5	7 mm
6	16,244 m
7	10,100 kg
8	0,022 kg
9	5,005 l
10	0,254 m

14-2

1	13,863 m
2	6,108 m
3	127,497 m
4	19,395 kg
5	4,243 ℓ
6	14,827 m
7	75,563 ℓ
8	0,773 ℓ
9	0,229 kg
10	4,268 759 m

14-3

1	37,023
2	0,014 668
3	0,005 12 m^2
4	264,660 48 m^2
5	1,759 262 kg•m

14-4

1	1,87
2	21,969 5
3	18,957
4	0,435 m
5	3,265 J /s
6	84,289 kg

ANSWERS, Chapter 15

15-1

1	122 mm
2	113 mm
3	108 mm
4	97 mm
5	67 mm
6	23 mm
7	53 mm
8	19 mm
9	35 mm
10	14 mm
11	18 mm
12	8 mm
13	5 mm
14	10 mm
15	7 mm
16	11 mm
17	5 mm
18	24,5 mm
19	10,5 mm
20	4 mm
21	14 mm
22	26 mm
23	12 mm
24	12,5 mm
25	7 mm
26	12 mm
27	6 mm
28	8 mm
29	100 mm
30	16 mm

15-1 contd.

31 6,5 mm
32 16 mm
33 12 mm
34 28 mm
35 13,5 mm
36 6,5 mm
37 18,5 mm
38 27 mm
39 13,5 mm
40 45,5 mm

15-2

1 450 mm
2 1700 mm
3 1640 mm
4 1110 mm
5 610 mm
6 1100 mm
7 33 in.
8 258,3 in.^{2}
9 840 mm
10 89 – 61 – 91 cm

15-3

1 13,8 m
2 3,9 m
3 9,9 m
4 6 m
5 6 m
6 2090 mm
7 219,5
8 no
9 13
10 *a,c,b*

15-4

1 5000 mm
2 47 ft 11 in.
3 $70\frac{7}{8}$ in.
4 70 in.
5 6 ft
6 50 km/h
7 80 km/h
8 35 mi/h
9 73,6 m
10 2,88 s

15-5

1 29 m
2 10,4 m
3 19,6 mm
4 8,67 km
5 3,7 m/s^2

ANSWERS, Chapter 16

1 180 m^2
2 135 m^2
3 1,35 ares
4 *a*) 6000 ft^2
   b) 0,138 acres
   c) 557 m^2
   d) 5,57 ares
5 *a*) $6,97 \times 10^6$ ft^2
   b) 160 acres
   c) $6,48 \times 10^5$ m^2
   d) 0,648 km^2
   e) $6,48 \times 10^3$ ares
   f) 64,8 hectares
6 1700 mm^2
7 507 mm^2
8 8,25 m^2
9 7854 mm^2
10 0,258 mm^2

ANSWERS, Chapter 17

1 *a*) $1{,}137 \times 10^{-3}$ m^3
b) $1{,}137 \times 10^{6}$ mm^3
c) 1,137 liters
2 42,9 mm^3
3 450 m^3
4 10^9
5 0,832 m^3
6 832 liters
7 *a*) 216×10^{3} mm^3
b) 216×10^{-6} m^3
c) $0{,}216$ liters $\times 10^{-3}$ liters
d) 0,132 cu. in.
8 *a*) 236×10^{6} mm^3
b) 236×10^{-3} m^3
c) 236 l
9 35,2 liters
10 4,54 liters
11 227 mℓ
12 474 mℓ
13 2,831 6 m^3; say 3 m^3

ANSWERS, Chapter 18

1 2,2 lb
2 397 g
3 8,9 oz
4 2204,6 lb
5 622 g
6 4,32 Mg
7 1,06 kg
8 530 kg
9 5 bags
10 1,342 kg
11 wood = 2648,7 kg
shavings = 1324,3 kg
chips = 1854,1 kg
sawdust = 2118,9 kg

ANSWERS, Chapter 19

1 *a*) $\frac{\pi^r}{3}$, $1{,}05^r$
b) $\frac{\pi^r}{4}$, $0{,}785^r$
c) $\frac{\pi^r}{9}$, $0{,}349^r$
d) $\frac{\pi^r}{2}$, $1{,}57^r$
e) $\frac{2\pi^r}{3}$, $2{,}09^r$
f) $\frac{5\pi^r}{4}$, $3{,}93^r$

2 *a*) 360°
b) 30°
c) 270°
d) 150°
e) 57,3°
f) 14,7°

3 *a*) 0,75°
b) 0,333°
c) 20,2°
d) 65,4°
e) 104,5°
f) 0,016 67°

4 *a*) 0,0100°
b) 0,001 39°
c) 0,505°
d) 5,109 7°
e) 81,322°
f) 225,927°

5 $\frac{4}{9}$ sr = 0,444 sr

6 1,39 sr

ANSWERS, Chapter 20

1 *a*) 0,333 h
b) 0,0833 h
c) 2,30 h
d) 5,67 h
e) 12,25 h
f) 0,016 7 h
2 *a*) 0,5 min
b) 0,917 min
c) 4,2 min
d) 15,467 min
e) 2,301 4 h
f) 15,594 h
3 \$27,57
4 \$42,34
5 3,6 h
6 4 h 43 min 18 s; **4,722 h**
7 03.20 h
8 16.40 h
9 9:20 p.m.
10 8,5 h

ANSWERS, Chapter 21

1 21°C
2 37°C
3 −40°
4 1 degC = 1,8 degF
5 −273,15°C
6 −153,15°C
7 298,15 K
8 283,15 K
9 68°F
10 287,33°F

ANSWERS, Chapter 22

1 240 N
2 2,75 m /s^2
3 134,6 kg
4 9,81 N
5 11,8 kg

ANSWERS, Chapter 23

1 104 mm
2 69,33 mm
3 17,85 mm
4 15,5 mm
5 12,3 mm
6 67 mm, 61 mm
7 7 mm
8 $l = 80$ mm, $h = 50$ mm
$w = 40$ mm
9 $l = 8$ cm $= 0{,}08$ m
$w = 4$ cm $= 0{,}04$ m
$h = 5$ cm $= 0{,}05$ m
10 2560 mm, 30 mm, 8 mm
11 1,22 m
12 0,07 m, 70 mm
13 0,0854 m^3
14 *a*) 3,6 m, 3600 mm
b) 20,79 m^2
c) 95,76 m^2
d) 18,975 m^3
15 *a*) **11,5 m^2**
b) **27,75 m^2**

ANSWERS, Chapter 24

1 a) 90 kelvins
 b) 90 deg C
 c) − 53,15 °C; 36,85°C

2 9,72 kW•h

3 Oil B

4 92,7 W

ANSWERS, Chapter 25

1 2,5 mA

2 0,2 MΩ

3 7,8 V

4 125 mW

5 231 W

6 4,16 kW•h

7 200 kHz

8 $C = \dfrac{0{,}885 \times 10^{-6}\epsilon_r A(N-1)}{t}$ pF

9 0,418 aF

10 2,57 kΩ

11 $L = \dfrac{25.4n^2r^2}{9r + 10l}\,\mu\text{H}$

12 106 μH

13 18,5 Ω

14 12 MHz

ANSWERS, Chapter 26

1 50 lm

2 0,25 sr

3 1,33 klux

4 2,55 klm

5 18 cd

Index